PREPARING FOR CALL CENTER INTERVIEWS

Published by
Lotus Press

PREPARING FOR CALL CENTER INTERVIEWS

Namrata Palta

Lotus PRESS
4263/3, Ansari Road,
Darya Ganj, New Delhi-110002

Lotus Press
4263/3, Ansari Road, Darya Ganj, New Delhi- 110002
Ph.: 32903912, 23280047 • E-mail : lotus_press@sify.com
www.lotuspress.co.in

PREPARING FOR CALL CENTER INTERVIEWS

ISBN: 81-8382-106-5
Reprint: 2009, 2010

Published by : **Lotus Press,** New Delhi
Printed at : Vishal Binding House, Noida

Preface

Everyday, we hear of people who want to join an international call center. However, they are not able to clear the rounds of the interview. To enable aspiring call center candidates to get through the interview in the first attempt itself, we have designed this book.

The book has sections, which focus on different interview rounds, held in call centers. The reader needs to go through each section and then form his answers based on the questions provided in the book. He also needs to understand the requirements of an international call center and prepare accordingly before facing an interview.

International call centers require a person to meet criteria, which is achievable if prepared for. If you are interested in joining an international call center, then work on your pronunciation and practice the interview questions along with this book. Make sure that you spend at least a couple of hours on this book everyday along with improving your accent.

Happy Learning & All the Best!

Contents

1
UNDERSTANDING CALL CENTERS

What is a Call center?

A **call centre** (UK) or **call center** (US) is a centralized office of a company that answers incoming telephone calls from customers.

A call center may be an office that makes outgoing telephone calls to customers (*telemarketing*). Such an office may also respond to letters, faxes, e-mails and similar written correspondence.

However the term **contact center** (UK) or **contact center** (US) is often applied when such multiple functions are blended in one office.

What Is Outsourcing?

The concept of taking internal company functions and paying an outside firm to handle them. Outsourcing is done to save money, improve quality, or free company resources for other activities. Outsourcing was first done in the data-processing industry and has spread to areas, including telemessaging and call centers. Outsourcing is the wave of the future.

What is Business Process Outsourcing (BPO)?

The concept of outsourcing started with Ross Perot when he founded Electronic Data Systems in 1962. EDS would tell a prospective client, "You are familiar with designing, manufacturing and selling furniture, but we're familiar with managing information technology. We can sell you the information technology you need, and you pay us monthly for the service with a minimum commitment of two to ten years.

BPO is the act of transferring some of an organization's repeated non-core and core business processes to an outside provider to achieve cost reductions while improving service quality. Because the processes are repeated and a long-term contract is used, outsourcing goes far beyond the use of consultants. If done well, BPO results in increasing shareholder value.

The main difference between BPO and more traditional IT outsourcing is that BPO offers companies a way of achieving transformational outcomes much more quickly.

In a typical BPO contract, a service provider takes over a specific corporate function. Effective BPO encompasses much more than just changing who is responsible for performing the process. In BPO, the outside provider not only takes on the responsibility to manage the function or business process, but also re-engineers the way the process has been traditionally done.

The next generation of Business Process Outsourcing has emerged as a priority for businesses looking to better options in managing their application portfolios.

The first wave offered low-cost, offshore development labor, but today firms are demanding new, less risky options for applications that are strategic, complex, or mission-critical, while still taking cost into consideration.

Outsourcing has moved from a niche technology management tool to a mainstream strategic weapon. Business Process Outsourcing leverages process driven efficiencies in terms of organizational excellence, responsiveness & branding, financial efficiency and customer relationship.

BPO is emerging as a powerful and flexible approach that business leaders can use to achieve a wide range of tactical and strategic aims. The most common business process that gets outsourced is call centers.

Call centers and Help Desks of many multi national and fortune 500 companies are being outsourced to low waged, English speaking countries such as Philippines and India.

Countries like India with vast IT human resources are also attracting outsourcing from American IT/Technology companies to outsource their IT Help Desks. Many of these help desks are state of the art with latest Help Desk software and help desk hardware with technical savvy IT graduates behind them answering your questions.

What is the difference between Domestic and International call centers?

Domestic call centers function within the domains of the country they exist in and international call centers are servicing clients from overseas.

Difference between Inbound and Outbound Call centers-Call dynamics

Types of calls are often divided into outbound and inbound.

Inbound calls are calls that are initiated by the customer to obtain information report a malfunction or ask for help.

This is substantially different from *outbound* calls where the agent initiates the call to a customer mostly with the aim to sell a product or a service to that customer.

The staff of the call center is often organized in tiers, with the first tier being largely unskilled workers who are trained to resolve issues using a simple script. If the first tier is unable to resolve an issue the issue is escalated to a more highly skilled second tier. In some cases, there may be third or higher tiers of support.

Classification of Call Centers

Types of Call Centers

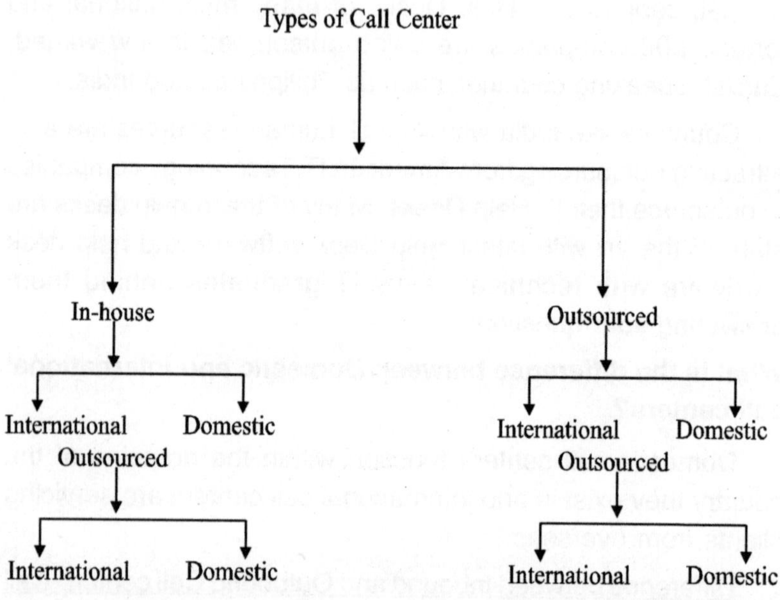

Some FAQs Relating To call center

What is a call center?

It's a functional area within an organization or an outsourced, separate facility exists solely to answer inbound or place outbound telephone calls. Usually refers to a sophisticated voice operations center that provides a full range of high-volume, inbound or outbound call handling services, including customer support, operator services, directory assistance, multilingual customer support, credit services, card services, inbound and outbound telemarketing, interactive voice response and web-based services.

How many call centers are there?

This is a hard question to answer, because the truth is nobody knows. A lot depends on how you define a center. Do you define it by the number of the people who can work there at any one time ("seats")? Or by kind of technology used in it? Or by basic function-some people don't count either dedicated inbound or outbound center, for their own strange reasons.

There are somewhere close to 140,000 call centers in the US and Canada. There are perhaps 20,000 more scattered around the world.

What's the difference between inbound and outbound centers?

Literally, an inbound center is one that handles calls coming in from outside, most often through toll free numbers. These calls are primarily service and support calls, and inbound sales.

An outbound center is one that does mainly outgoing telemarketing.

Inbound is the biggest component of the call center traffic these days, though perversely, outbound represents the area of largest projected growth in the next few years.

In truth, the majority of centers contain some element of both inbound and outbound.

What are the origins of the call center industry?

Some time around thirty years ago, the travel and hospitality industry began to centralize their reservation centers into what we would recognize now as huge call centers. This happened at around the time the first large-scale high-volume premise-based telephone switches became available.

Banks have also used them since the 1970s at least, and

later in that decade, with the rise of the catalog shopping movement and outbound telemarketing, call centers became a staple within many industries. Each industry, however, had is own way of operating centers, its own standards for quality, and its own preferred technologies. This balkanization of the industry persisted until early in the 1990s, when call center managers became more recognized as having a consistent set of skills and an operational knowledge.

In 1984, Business week wrote the following:

"Much of the impetus for establishing such centers was generated by Technical Assistance Research Programs Inc. (TRAP), a Washington consulting firm hired by the White House to assess service in U.S. TRAP's findings were eye-opening, suggesting that corporations were inadvertently losing many of their customers...Companies as diverse as GE, AmEx, GM's Buick, & Coors mentioned the TRAP studies when discussing their service operations. To make their companies more accessible, they began to advertise toll-free lines. But more

Than phone lines are needed, as several companies are discovering. The 800 numbers can be an added source of customer frustration unless they provide fast & efficient response. A number of companies that have installed the lines do not cover them continuously—or at the times when customers would be most likely to use them..."

And so began in earnest the debate about call center: cost control versus service improvement.

What does a call center manager actually do?

A call center manager's job consists of ensuring the continued daily operation of the center. He or she is responsible for setting service standard for the center (i.e., how long before calls get answered, on average, or how many contacts per hour outbound agents are required to make).

The manager is responsible for solving intraday problems, arranging for having the correct numbers of agents in the right number of seats, & for making sure that the center operates with enough of the right kinds of technologies to make the customer experience a positive one. The call center manager's job is to put out fires all day long, & it's truly impossible one.

(Why are call centers constantly called other things, like "contact centers," "customer care centers," "support centers," &ugly constructions like "customer contact zones" or "multi media access centers"?)

These terms are the industry's ongoing attempt to come to grips with changes in what a call center is supposed to do. Now that you have call centers that answer e-mail, engage in live Internet chat session with customers & sometimes even transmit live video, people are looking for a broader definition to replace "call center." Rest assured, they all mean basically the same thing.

Understanding the Role of a Customer Care Executive

CUSTOMER CARE EXECUTIVE is a person who interacts with customer in call center. They are the first point of contact with the current and potential customers. Appropriate training of staff can have enormous impact on the customer loyalty and satisfaction and the Company's can minimize risk of losing valued customer.

Attributes of a Customer Care Executive:

- Loves interacting with people: Since the position of a CUSTOMER CARE EXECUTIVE requires talking to people, person should have the basic trait to be able to enjoy the work and should be able to communicate confidently and effectively.
- Maintains high standard of professional ethics: Agent is in charge of confidential information pertaining to the client and the customer, which should not be disclosed.

- Is a team member: Contact centers are grooming grounds for leaders and every leader starts the journey by being a competent team player. Agent should strive to be responsive and sensitive to peers.
- Have good communicational skills: As the agent is in direct interaction with the customer, he needs to have good communicational skills so that he is able to answer the queries of the customer with conviction and accuracy.
- Respects individual rights: Agent respects individual rights to privacy and uninterrupted leisure time with respect to calling in odd hours.
- Identifies with the client values: Agent represents client while interacting with customer when they are not direct employees of client. There exists relation between client and agents in such a case, as customer believes he is interacting with somebody at clients end.
- Is a positive thinker and high on learning: Call Center is a dynamic industry. Agent should be able to analyze trends and address them so as to affect nature in positive way. One should have keen learning attitude. Sharing of the best practices and imbibing them to achieve higher productivity levels.

Growth of a Customer Care Executive

CEO

Operations Manager

Assistant Manager (Operations)

Shift Supervisor

Team Leader

Customer Care Executive

(Career graph of agent)

Contact centers offer excellent growth opportunities. This industry in India owes its existence to Internet and telecommunication because which cost of transferring voice data is negligible.

India is becoming preferred global destination for companies to outsource their operations because of the following advantages:

- Availability of well educated, English speaking manpower.
- Low infrastructure and manpower cost.
- Stable political environment that makes investment safe.
- Favorable time difference with developed economies.

By outsourcing IT Enabled Services, large companies are not only achieving significant benefits in cost but also creating platform for new business.

Introduction to Customer Service

"A Customer is someone who buys regularly. Hence he/ she who returns to buy whenever the need arises could be called a customer. So anyone at any time and at any place could be a customer. All of us are customers and service providers all the time. These people may be internal customer or external customers".

External Customers are those who buy our products and services; whereas Internal Customers are individuals or departments, who use our output as their input to create value for our external customers.

During the day you may contact somebody in your own organization for example, the Finance department for your bills reimbursements. Therefore during the day one could be several times a service provider and several times a customer.

It is important to understand the role of a service provider because as a customer if you expect good service then your customer also expects the same from you.

Customer Expectations:

Responsiveness

It refers to the help desk's ability to be available when customer needs help

This makes it easy for customers to contact the help desk.

Caring Attitude

A caring, positive and helpful attitude goes a long way to keeping customers satisfied. Although the help desk may not always be able to give customers exactly what they want, when they want it, there is always something the help desk can do. In other words, the help desk must be willing to assist customers in any way it can.

Fast Service

When service is slowed for any reason, you should give a general apology

"I'm sorry about this delay", or

"I know you expected faster service. Please accept my apologies".

Courteous Service

Other key words are: Please, You May, if you please and if you don't mind

"Thank you", or "Thanks for calling"

Consistent Service

It means the customer can depend on a high level of service every time he calls

Someone Who Will listen

Customer expects you to listen to their needs

Knowledgeable Representative

They want to make their inquiry and get immediate and

accurate response.

"There is little difference in people but that little difference makes a big difference. The little difference is Attitude. The big difference is whether it is positive or negative".

Commitment

Customers want you to make a specific commitment and follow through on it.

Call Centers Terminology

Call center terminology can be an intimidating group of terms and abbreviations so we've created a resource of commonly used terms specific to the industry to help you.

Call Center

Call center is a general industry term referring to many different types of call center facilities. Typically this is a center that handles one or many different types of services such as help desk, lead generation, emergency response, telephone answering services, outbound telemarketing, direct response, and customer service, among others.

Abandoned Call

A call where the caller disconnects before having reached an agent.

Agent

A call center staff member handling outgoing or incoming calls. There are a variety of names that a call center agent may be known as such as CUSTOMER CARE EXECUTIVE (Customer Service Representative), operator, or communicator, to name just a few.

Ani

A Call center uses ANI (Automatic Number Identification) to keep track of phone numbers people calling in to the call center. This is particularly useful in reporting crank callers or threats from callers to the appropriate authorities. Caller ID is the phone company version of ANI.

Answering Service

Call center service focused on taking calls on behalf of a person or business during break periods or after hours. Typically, a simple message is taken and relayed back to the client enlisting the call center answering service.

Auto Answer Greeting

Pre-recorded customized greeting played at the beginning of a call answered by the system. This greeting can be created by a call center staff member, or directly by the call center client, and played to the caller before the call is routed to an agent.

Call Distribution

In a call center environment where there are many different types of calls, it is important to ensure that each of the calls goes to the most appropriate call center agent to handle the specifics involved. This is set up by using a call distribution list, which tells the telephony software which agent can handle the call type.

Direct Response

Direct response is the process of measuring the results of direct media. This can be anything from a mail drop to television spots (DRTV), and everything in between. A contact number, usually a toll free number, is provided via the media source and people call in regarding the reason of the direct marketing. Types of calls that fall under the direct response umbrella include reservation bookings, lead capture, surveys, coupon redemption, catalog order entry, dealer locate, fundraising and donation lines, information and referral, mail order fulfillment, market research, merchant processing, and many more.

Dispatch

Dispatch calls are simply the calls that require a call center agent to take a message directly from the caller and relay the information to the person who the message was intended for. Commonly, dispatch calls are sent via text to a pager or as a text message to a cellular phone, or any other means of electronic text transfer.

DNIS

Dialed Number Identification Service. A string of numbers passed from the telephone network to identify what number the caller dialed. The system can then report on the number dialed to track various user/client defined specifics.

Emergency Response

Emergency Response calls generally refer to calls requiring specific and immediate handling by the call center agent. The caller supplies the reason of their call and the call center agent escalates the information to the appropriate person, as supplied by the client, to relay the information to. Usually, the client supplies the contact information for an on-call person, and several backup on call people in the event the scheduled on call person cannot be reached to handle the situation.

Handling Time

The time a call center agent spends connected to the caller combined with the time the call center agent spends handling the requirements of the call after the caller has disconnected.

Help Desk

Help Desk calls coming in to a call center are typically referring to calls for product usage and support or problems. Help desk calls often refer to but are not limited to computer software packages or hardware configurations.

Inbound

Call center services are generally broken down into two basic categories-inbound and outbound. Inbound calls are the calls received by the call center on behalf of a client. They consist of help desk calls, direct marketing calls, emergency response calls, answering service calls, and any number of other types of calls where the caller dials a number that connects to the call center.

IVR

Interactive Voice Response. IVR is often used to provide an interactive menu message, providing the caller with keypad options for directories of information or call direction. It is a common feature used in larger business to direct callers to the department that best suits their needs. IVR can also be used to log the answers to survey questions by instructing the caller to press specific buttons on the telephone keypad which would correspond with programmed answers.

Monitoring

This is a particularly useful way to ensure the quality of calls handled by call center agents. Constant evaluation of how a client's account is handled greatly reduces the chance of mishandled situations. In fact, for DRTV or direct response clients, it more often results in higher conversions of information only to specific desired call types.

Outbound

Call center services are generally broken down into two basic categories-outbound and inbound. Outbound calls are the calls made by call center agents on behalf of the client. They consist of various types of telemarketing such as not for profit and fundraising calls, contact list cleaning and updating, verification services, and selling products or services to name a few.

Paging

Paging is a common method of delivering a message to our client. It can be a text message being sent to a pager unit or cell phone, or it could be faxed or even emailed.

Pre-announce

A pre-announce is a recorded informational or instructional message that is played to people who call in on a client's account. The line is answered by the system and the message is played before routing the call to a live call center agent.

Predictive Dialing

Once again, falling under the category of Outbound Call Center Services, predictive dialing is a process whereby a database of phone numbers is loaded into the system and is then automatically dialed. Once a call connection is made, the system sends the live call to an outbound call center agent to handle. When a predictive dialer detects there is no answer, or the line is busy or an answering machine, it disconnects and puts the number back into the queue to be dialed again.

Queue

The queue is where the system holds callers who are not connected with a live call center agent. When someone calls in and there are no live agents available to take the call, the call is then placed in queue until an agent becomes available to take the call.

Service Level

Service level is the percentage of incoming calls that are answered live in an established reasonable amount of time. For example, 96 out of 100 calls are answered in 30 seconds or less would result in a 96% Service level.

Talk Time

The time that a call center agent spends on the phone with a caller is known as talk time. Talk time includes everything from the time the agent greets the caller, to the time they make their closing statement and disconnects the call.

Toll-Free Numbers

Toll free numbers allow callers to reach a call center no

matter of location, without incurring long distance charges. Typically, toll free numbers begin with 800, 888, 866, and 877. Many companies have vanity toll-free numbers such as 1-888-RESPONSE (1-888-737-7667).

2
PREPARING FOR THE INTERVIEW

Steps in preparing for an interview

Step 1. Finding Job Opportunities

Step 2. Writing your cover letter

Step 3: Writing your resume

Step 4. Preparing answers for important questions

Step 5: The Interview

1. Writing your covering Letter

A cover letter is the first impression an employer has of you, you need to stimulate the reader to read your resume.

Your cover letter should demonstrate your understanding of the position and the organisation you have applied for and how your skills and experience match the position.

Structure of a Cover Letter

Your name

Address

Phone

Email

Date

Contact Person's Name

Contact Person's Position

Company Name

Company Address (Don't forget this.)

Re: Name of position *or* Application for ...

Dear Ms Jones, (Avoid Dear Sir/Madam)

First paragraph

I would like to apply for the above position advertised in .... on .... (date) and why you want that job (i.e. why that industry and why that company, make sure this highlights how you will meet their needs, not exclusively your own.)

Second / Third paragraph

Outline the skills (technical and social), experience, training and achievements that are relevant to the selection criteria or the job description. These skills may have been gained through your education/course, your work experience and your extra curricula activities.

Fourth paragraph

Tell them why you are the best fit for the company and how your skills will meet their needs. (Tip: you could summarise the 3 best things about you that make you most competitive for the job and how these things will allow you to make an impact/get results in the role.)

Closing

Thank them for considering your application, tell them that you have attached a copy of your resume, and when you would be available for an interview. Provide your contact details (Phone / Fax / Email)

Yours sincerely,

Your signature

Print name

2. Resume Preparation

A resume is a marketing tool that lets an employer know what you have done, the experience that you have gained and what you can offer the organisation.

Your resume should match your skills and experience to the needs of the position/organisation.

Resume Success Checklist

- Check this list before completing and sending your resume
- What is the overall role of this job within the department and within the organisation?
- What skills are listed as necessary for the job? These could be generic or general skills such as communication or teamwork skills or they could be skills related to a particular professional area, such as taxation planning or writing press releases.
- What other skills do you think are important for the job, but are not listed?
- What experience is listed as necessary for the job?
- What other experience do you have that could be useful for the job, but is not listed.
- What educational qualifications are listed as necessary for the job?
- What other educational qualifications could be useful? Why?
- What on the job or off the job training have you done that is relevant to the job?
- What other self directed or self managed learning have you done that has developed skills and knowledge relevant to the job?
- What personal qualities are listed as necessary for the job?
- What other personal qualities, that are not listed, do you

think are important for the job? Think about the context of the job and the organisation and what the important needs and drivers are, both now and for the future. This will help you read between the lines to discover other important skills and qualities the organisation needs.

- It is a good idea to highlight the closing date and the contact person's details as this is usually where you send your application and you can also contact this person to find out more information before writing your application. Whilst the information is fresh in your mind, note the reasons you want this job and why you think you would be good at it. This information will be helpful later when you are writing your cover letter and your career objective in the resume.
- What kinds of results or outcomes is this position expected to achieve?
- What kinds of opportunities, success and achievements could be available in this job for someone who had initiative, drive and capability?
- What skills and personal qualities would make it easier to realise these opportunities? (Thinking about these questions is important as they will help you to think like the employer and will help guide the skills, knowledge and accomplishments that you highlight in your resume and letter.
- What kinds of challenges, barriers, difficulties and problems are likely to emerge and face the person doing this job?
- What skills and personal qualities would make it easier to face these challenges successfully? (It is important to have a good understanding of the nature of the challenges and difficulties involved in a role as particular skills and knowledge and personal qualities will be important for meeting and overcoming these challenges. If you know what

these are, you can take the opportunity to highlight your abilities in these areas in your resume and therefore reinforce your suitability for the role.)

Resume Layout

- There is no one right way to do a resume, just certain principles or guidelines.
- There are usually 2 culls for a resume – 20 seconds and then an in-depth review.
- Use labeling, clear formatting and alignment and white space to ensure your valuable information is noticed.
- Include enough detail and concrete examples for the second cull as well!
- 2-3 pages are the average for graduate resumes (although shorter or longer are also OK).
- If a length or format is stipulated – stick to it.
- Use only one plain font – Times New Roman, Arial, Helvetica, Century Schoolbook
- Use a combination of bullets and short paragraphs.
- Avoid italics and underline (they don't fax or scan into database well) and make it more difficult to read, as does the use of too many brackets.
- One typo can reduce your chances of success by 70% - edit furiously
- Don't over capitalise or over bold (e.g. 2 or 3 lines in a paragraph)
- Be consistent in formatting. Put dates, titles, full stops etc in similar places throughout your resume.
- Format each section similarly if possible to help the reader digest your information.

- Be consistent with your language throughout your resume.
- Use active verbs to indicate achievements.
- There is no one right way to set out a resume. Think about what the employer will value most.
- Make the resume is as simple and easy to read as possible.
- Make sure that your resume is well presented and well organised, so that an employer can immediately see the most relevant information - remember clear headings and lots of white space.
- Balance the resume as to what the employer will consider most important / relevant (eg don't spent a page describing high school achievements and one line describing your degree)

Presentation

Make sure that your application is typed and professionally presented.

Remember that this is often the first contact with an employer, and therefore spelling and grammatical errors are unacceptable and may result in your application not being considered. It is very hard to convince someone that you have strong attention to detail if your application is full of mistakes.

Paper

Ordinary, white, A4 paper is perfectly acceptable, or you can be more daring and use better quality, higher grade paper, as is often used in letterheads.

Print on dark paper may be difficult to read and will not photocopy well.

Sample Interview Questions

Introductory / Background Questions

These questions are designed to find out a little bit more

about you. It is a chance for you to explain the major aspects of your resume, what you've done, your attitudes and values, and the choices you've made in your life and most importantly, why you made those choices. It is often a good idea when answering these types of questions to start with the most recent information and work backwards and also to start with a brief synopsis. Remember, that you need to be informative without going into so much detail that you bore them. It is also worth noting that if they ask you to expand on what you are saying, they are asking because they are interested and not because you've necessarily done something wrong.

Tell me about yourself

What are your greatest strengths and weaknesses?

How would you describe yourself?

Tell me about any time when you weren't pleased with your performance. What did you do about it?

Tell me about a challenging situation you have faced? What was the situation and how did you cope with it?

Objectives

This is a somewhat difficult area on which to answer questions. It is difficult because you need to demonstrate to the interviewer that you have thought about what it is that you want from work and what you hope to achieve. At the same time you need to assure them that you are flexible and would be prepared to work in a range of areas. Therefore, you need to tell them about those areas that are of most interest to you, as well as reassuring them that you are flexible and willing to explore new opportunities.

- What are your long range and short range goals and objectives? How are you preparing yourself to achieve them?
- What do you see yourself doing five years from now?

- Why did you choose the career for which you are preparing?
- What do you see yourself doing in five years' time? Do you have any long range ambitions?
- What are you looking for in a job?
- What do you expect from your initial graduate position?
- In what kinds of positions are you most interested?
- Have your thought about any other careers to which you may be more suited?
- What job would you look for if you could not find a job in this field?

Job Interests

Employers are often keen to find out what your major areas of interest are so that they can better assess how well you will fit into their organisation, both now as well as your potential to develop with the organisation in the future.

- In what kinds of positions are you most interested?
- What are the most important considerations for you in choosing a job?
- What is more important to you, the money or the type of job?
- Do you have a geographical preference? Why?
- Will you relocate? Does relocation bother you?
- Are you willing to travel?
- Where would you like to work?
- How much do you expect to be paid?
- What other applications have you made?
- Are your over qualified?
- What's on your shopping list?

- You do realise, don't you, that this is a junior position?
- What sort of training/induction would you expect to be involved in initially?
- Most jobs have their frustrations and problems. Describe some that have been particularly dissatisfying to you.
- What have been the most satisfying aspects of jobs that

Other Interests

With these questions employers are trying to find out more about the types of things in which you are interested, so that they can gain a better understanding of you, and thereby better assess how well you are going to fit in with the company, both now and in the future. There are no right answers to these questions but it is important to think about how your interests and goals fit in with what the company does and how it hopes to develop in the future.

- Where can you be found when you are not working (or studying)?
- Who is your favourite novelist?
- Do you like/dislike football?

Educational Questions

With these questions interviewers hope to gain more of an insight into what it is that you have studied so that they can better assess how well your studies meet their needs. Most importantly, they are interested in the reasoning behind why you chose the course that you did and what you think that you got out of it. You will find that these questions are likely to focus on both the technical as well as the generic skills that you develop through your course.

- Why did you choose to major in ...?
- What led you to choose your field or major of study?

- What university subjects did you like best? Least? Why?
- Do you have plans for continued study? An advanced degree?
- What courses have you taken that prepared you for this position?
- What part of your education do you see as being most relevant to this position?
- Is your academic performance any indication of how you will do this job?
- What do you like about your discipline of study?
- How does your academic performance compare with your colleagues and what factors have influenced it?
- What parts of your education do you see being most relevant to this job?

Achievements

What two or three accomplishments have given you the most satisfaction? Why?

Is there an achievement of which you are particularly proud? What is it and why is it significant?

Work Experience

- In what part time or summer jobs have you been most interested? Why?
- Do you have any vacation employment or previous experience in this or a related field?
- What type of work experience have you had and how do you feel that it has helped prepare you for a graduate position?
- What has been the value of your previous work experience?
- Your Contribution to the Organisation

- Why should I hire you?
- In what ways do you think you can make a contribution to our company?
- Can you see any conflicts of interest between your career and that of your partner?
- Why should we give this job to you?
- Is there anything else we should take into account to help us make a decision?
- So far you have not really convinced me I should give you this job.
- What particular skills and qualities do you bring to the workforce? What other skills would you like to develop in the future?

Commitment to the Company

An employer is interested to know how you think that you might fit into their organisation.

- Why did you decide to seek a position with this company?
- What do you know about our company?
- Why did you apply to this organisation?
- Are you prepared to perform duties that may not be part of your routine?

Commerciality

You mentioned you had a part time job in retail/hospitality, what is the business environment of the company? What are the goals of your team?

- What do you understand is the current climate of the ____ industry?

Approach to Work

- Employers are often interested in knowing how you manage

different responsibilities, particularly how you prioritize your time.

- How do you work under pressure?
- How do you balance work priorities and those of your personal and family life?
- Describe the relationship that should exit between a supervisor and those reporting to him or her.
- In what kind of work environment are you most comfortable? What two or three things are most important to you in your job?
- How do you get on with other people?
- How do you plan your day?
- Tell me about a time when, despite careful planning, things got out of hand.
- Here is a problem we faced recently ... How would your have dealt with this?
- Can you handle pressure?
- How would you deal with a problem you had not encountered before?
- How high do you set your standards?
- Have your ever been responsible for the generation of a new idea or suggestion?
- Describe and explain.
- Describe how you have coped with a new or difficult situation.
- Have you ever had to persuade people to your way of thinking?

Teamwork

Have you been in any situations where you have been

required to perform as part of a team? What was the situation, what part did you play in the team and what was the outcome of the exercise?

Outline a time you were in a team situation. What was your role? What went well/not so well? How did you contribute to the enjoyment of the team? How did you confront a difficult team member?

- How do you establish working relationships with new people?
- What kind of people do you like to work with?
- Have you ever had to stifle your normal behaviour to get along with someone?
- How do you define a conducive work atmosphere?
- What do you feel are the key ingredients to a successful team?
- What are some of the problems that can occur within team

Your Turn to Ask

Remember, interviews are a two way process. During the interview, the employer will be trying to find out as much as they can about you, but equally, it is your chance to find out as much as you can about them. So ask questions! If there are particular things that you are looking for in a job and they haven't been mentioned into the interview, ask about them.

- Why is this position open?
- What are the key requirements for the position?
- How is performance measured? (Goals should be clearly defined so measurements have some objectivity.)
- How tough is your company's opposition? (You should know from your research who their major competitors are.)
- Are there opportunities that are unique to this job within the organisation?

- Do internal promotions meet your growth demands for management?
- Do you feel I fit the requirements for the position? From your perspective, in what areas do you see that there are gaps?
- Are there any areas you would like me to expand upon?
- What are the company's plans for the future?
- What would a successful person expect to be earning in their fourth or fifth year?
- What sort of induction and training do you normally provide?
- What do you like most and least about working at ____?
- Why did you choose to work at ____?
- I have researched the organisational culture and what appeals to me is ..... Can you tell me about any social or sporting events that are organised by staff?
- Does the organisation interact with the local community (e.g. provide pro bono work)?

Most Important Interview Questions

1. **Tell me about yourself:** - The most often asked question in interviews. You need to have a short statement prepared in your mind. Be careful that it does not sound rehearsed. Limit it to work-related items unless instructed otherwise. Talk about things you have done and jobs you have held that relate to the position you are interviewing for. Start with the item farthest back and work up to the present.
2. **Why did you leave your last job?** - Stay positive regardless of the circumstances. Never refer to a major problem with management and never speak ill of supervisors, co-workers or the organization. If you do, you will be the one looking bad. Keep smiling and talk about

leaving for a positive reason such as an opportunity, a chance to do something special or other forward-looking reasons.

3. **What experience do you have in this field?** - Speak about specifics that relate to the position you are applying for. If you do not have specific experience, get as close as you can.
4. **Do you consider yourself successful?** - You should always answer yes and briefly explain why. A good explanation is that you have set goals, and you have met some and are on track to achieve the others.
5. **What do co-workers say about you?** - Be prepared with a quote or two from co-workers. Either a specific statement or a paraphrase will work. Jill Clark, a co-worker at Smith Company, always said I was the hardest workers she had ever known. It is as powerful as Jill having said it at the interview herself.
6. **What do you know about this organization?** - This question is one reason to do some research on the organization before the interview. Find out where they have been and where they are going. What are the current issues and who are the major players?
7. **What have you done to improve your knowledge in the last year?** - Try to include improvement activities that relate to the job. A wide variety of activities can be mentioned as positive self-improvement. Have some good ones handy to mention.
8. **Are you applying for other jobs?** - Be honest but do not spend a lot of time in this area. Keep the focus on this job and what you can do for this organization. Anything else is a distraction.
9. **Why do you want to work for this organization?** - This

may take some thought and certainly, should be based on the research you have done on the organization. Sincerity is extremely important here and will easily be sensed. Relate it to your long-term career goals.

10. **Do you know anyone who works for us?** - Be aware of the policy on relatives working for the organization. This can affect your answer even though they asked about friends not relatives. Be careful to mention a friend only if they are well thought of.

11. **What kind of salary do you need?** - A loaded question. A nasty little game that you will probably lose if you answer first. So, do not answer it. Instead, say something like, That's a tough question. Can you tell me the range for this position? In most cases, the interviewer, taken off guard, will tell you. If not, say that it can depend on the details of the job. Then give a wide range.

12. **Are you a team player?** - You are, of course, a team player. Be sure to have examples ready. Specifics that show you often perform for the good of the team rather than for yourself are good evidence of your team attitude. Do not brag, just say it in a matter-of-fact tone. This is a key point.

13. **How long would you expect to work for us if hired?** - Specifics here are not good. Something like this should work: I'd like it to be a long time. Or As long as we both feel I'm doing a good job.

14. **Have you ever had to fire anyone? How did you feel about that?** - This is serious. Do not make light of it or in any way seem like you like to fire people. At the same time, you will do it when it is the right thing to do. When it comes to the organization versus the individual who has created a harmful situation, you will protect the organization. Remember firing is not the same as layoff or reduction in force.

15. **What is your philosophy towards work?** - The interviewer is not looking for a long or flowery dissertation here. Do you have strong feelings that the job gets done? Yes. That's the type of answer that works best here. Short and positive, showing a benefit to the organization.
16. **If you had enough money to retire right now, would you?** - Answer yes if you would. But since you need to work, this is the type of work you prefer. Do not say yes if you do not mean it.
17. **Have you ever been asked to leave a position?** - If you have not, say no. If you have, be honest, brief and avoid saying negative things about the people or organization involved.
18. **Explain how you would be an asset to this organization** - You should be anxious for this question. It gives you a chance to highlight your best points as they relate to the position being discussed. Give a little advance thought to this relationship.
19. **Why should we hire you?** - Point out how your assets meet what the organization needs. Do not mention any other candidates to make a comparison.
20. **Tell me about a suggestion you have made** - Have a good one ready. Be sure and use a suggestion that was accepted and was then considered successful. One related to the type of work applied for is a real plus.
21. **What irritates you about co-workers?** - This is a trap question. Think real hard but fail to come up with anything that irritates you. A short statement that you seem to get along with folks is great.
22. **What is your greatest strength?** - Numerous answers are good, just stay positive. A few good examples: Your ability to prioritize, Your problem-solving skills, Your ability

to work under pressure, Your ability to focus on projects, Your professional expertise, Your leadership skills, Your positive attitude .

23. **Tell me about your dream job.** - Stay away from a specific job. You cannot win. If you say the job you are contending for is it, you strain credibility. If you say another job is it, you plant the suspicion that you will be dissatisfied with this position if hired. The best is to stay genetic and say something like: A job where I love the work, like the people, can contribute and can't wait to get to work.

24. **Why do you think you would do well at this job?** - Give several reasons and include skills, experience and interest.

25. **What kind of person would you refuse to work with?** - Do not be trivial. It would take disloyalty to the organization, violence or lawbreaking to get you to object. Minor objections will label you as a whiner.

26. **What is more important to you: the money or the work?** - Money is always important, but the work is the most important. There is no better answer.

27. **What would your previous supervisor say your strongest point is?** - There are numerous good possibilities: Loyalty, Energy, Positive attitude, Leadership, Team player, Expertise, Initiative, Patience, Hard work, Creativity, Problem solver

28. **Tell me about a problem you had with a supervisor** - Biggest trap of all. This is a test to see if you will speak ill of your boss. If you fall for it and tell about a problem with a former boss, you may well below the interview right there. Stay positive and develop a poor memory about any trouble with a supervisor.

29. **What has disappointed you about a job?** - Don't get trivial or negative. Safe areas are few but can include: Not enough

of a challenge. You were laid off in a reduction Company did not win a contract, which would have given you more responsibility.

30. **Tell me about your ability to work under pressure.** - You may say that you thrive under certain types of pressure. Give an example that relates to the type of position applied for.

31. **Do your skills match this job or another job more closely?** - Probably this one. Do not give fuel to the suspicion that you may want another job more than this one.

32. **What motivates you to do your best on the job?** - This is a personal trait that only you can say, but good examples are: Challenge, Achievement, Recognition

33. **Are you willing to work overtime? Nights? Weekends?** - This is up to you. Be totally honest.

34. **How would you know you were successful on this job?** - Several ways are good measures: You set high standards for yourself and meet them. Your outcomes are a success.Your boss tell you that you are successful

35. **Would you be willing to relocate if required?** - You should be clear on this with your family prior to the interview if you think there is a chance it may come up. Do not say yes just to get the job if the real answer is no. This can create a lot of problems later on in your career. Be honest at this point and save yourself future grief.

36. **Are you willing to put the interests of the organization ahead of your own?** - This is a straight loyalty and dedication question. Do not worry about the deep ethical and philosophical implications. Just say yes.

37. **Describe your management style.** - Try to avoid labels. Some of the more common labels, like progressive, salesman or consensus, can have several meanings or

descriptions depending on which management expert you listen to. The situational style is safe, because it says you will manage according to the situation, instead of one size fits all.

38. **What have you learned from mistakes on the job?** - Here you have to come up with something or you strain credibility. Make it small, well intentioned mistake with a positive lesson learned. An example would be working too far ahead of colleagues on a project and thus throwing coordination off.

39. **Do you have any blind spots?** - Trick question. If you know about blind spots, they are no longer blind spots. Do not reveal any personal areas of concern here. Let them do their own discovery on your bad points. Do not hand it to them.

40. **If you were hiring a person for this job, what would you look for?** - Be careful to mention traits that are needed and that you have.

41. **Do you think you are overqualified for this position?** - Regardless of your qualifications, state that you are very well qualified for the position.

42. **How do you propose to compensate for your lack of experience?** - First, if you have experience that the interviewer does not know about, bring that up: Then, point out (if true) that you are a hard working quick learner.

43. **What qualities do you look for in a boss?** - Be generic and positive. Safe qualities are knowledgeable, a sense of humor, fair, loyal to subordinates and holder of high standards. All bosses think they have these traits.

44. **Tell me about a time when you helped resolve a dispute between others.** - Pick a specific incident. Concentrate on your problem solving technique and not the dispute you settled.

45. **What position do you prefer on a team working on a project?** - Be honest. If you are comfortable in different roles, point that out.

46. **Describe your work ethic.** - Emphasize benefits to the organization. Things like, determination to get the job done and work hard but enjoy your work are good.

47. **What has been your biggest professional disappointment?** - Be sure that you refer to something that was beyond your control. Show acceptance and no negative feelings.

48. **Tell me about the most fun you have had on the job.** - Talk about having fun by accomplishing something for the organization.

49. **Do you have any questions for me?** - Always have some questions prepared. Questions prepared where you will be an asset to the organization are good. How soon will I be able to be productive? and What type of projects will I be able to assist on? are examples.

50. Why do you want to work here?

51. Why should we hire YOU?

52. What can you do for us that someone else cannot?

53. Why did you leave your last job?

54. Have you ever been fired or forced to resign?

55. Can you explain this gap in your employment history?

56. What have you been doing for the "X" months since you were last employed?

57. Why have you had so many jobs in such a short period of time?

58. What have you learned from your mistakes?

59. Have you ever had difficulty with a supervisor? With a

coworker?

60. How did you resolve the conflict?

61. Describe some times when you were not very satisfied or pleased with your performance. What did you do about it?

62. Give me an example of a problem you faced on the job, and tell me how you solved it.

63. What do you consider to be your greatest strengths and weaknesses?

64. What are your long-range career objectives, and what steps have you taken toward obtaining them?

65. Give me an example of an important goal you had to set and tell me about your progress in reaching that goal.

66. What was the best decision you ever made?

67. What was your role in your department's most recent success?

68. Describe a time when you were faced with problems or stresses at work that tested your coping skills. What did you do?

69. How do you deal with competition? Are you a competitive person?

70. What would you consider an ideal work environment?

71. How well do you work with people? Do you prefer working alone or in teams?

72. Describe a situation when working with a team produced more successful results than if you had completed the project on your own.

73. Tell me about a time you had to gain the cooperation of a group over which you had little or no authority. What did you do?

74. What do you do when people disagree with your ideas? Describe a situation.

75. Describe a situation where you found yourself dealing with someone who didn't like you. How did you handle it?

76. Do you consider yourself to be a leader? What are the attributes of a good leader? Give me an example of when you had to show good leadership.

77. Tell me about a time when you had to use your presentation skills to influence someone's opinion.

78. Can you tell me about an important written document you were required to complete?

79. Describe the most creative thing you have ever done.

80. What motivates you to go the extra mile on a project or job?

81. What does it mean to be successful? According to your definition, how successful have you been so far?

82. Give me an example of a time when you tried to accomplish something and failed. Were you discouraged by this? What did you do about it?

83. What do you really want to do in life? What do you see yourself doing five years from now? Ten years from now?

84. If you had six months ahead with no obligations and no financial constraints, what would you do?

85. You don't have enough or the right kind of experience. You may be overqualified or too experienced for the position. What are your expectations regarding promotions and salary increases?

86. How much do you expect if we offer this position to you?

Introduction at the Interview

What's the most important thing you can do before an interview?

Preparation.

And that's what this interviewing tutorial is all about. Giving you the tools you need to be better prepared for all aspects of interviewing for a job — the good and the bad; the types of job interviews; handling typical interview questions; and much more.

What is chemistry — and why is it important in a job interview?

The chemistry or rapport between you and the interviewer.

Do you "click" with the interviewer?

If not, there's a good chance you won't get the job.

It's depressing because there's not a lot you can do to improve chemistry.

How long will the interviewer take to decide whether you are a good fit with the company?

The interviewer will probably decide whether you are a good fit with the company in the first five minutes of the interview.

How can you prepare for the interview BEFORE you go in to meet the employer?

- Figure out as much as possible about the employer's needs.
- Find out more than the average interviewee about the company and the hiring manager's concerns.
- Compose answers to frequently asked interview questions. Check out the common interview questions
- Rehearse answers to those questions.
- If you have an opportunity to do practice interview, such a mock interview, do it.

- DO NOT rehearse canned answers out of books. Interviewers have heard them all.
- Transferable Skills: Be sure to point out your transferable skills, with examples that are not always evident from your work experience or education.

Strategic Portrayal of Transferable Skills

One of the most important concepts you will ever encounter in the job-hunting process is that of transferable job skills. The deft use of transferable skills should pervade your job search and be a key factor in your resume, cover letter, and interview strategies.

What are transferable skills?

Simply put, they are skills you have acquired during any activity in your life — jobs, classes, projects, parenting, hobbies, sports, virtually anything — that are transferable and applicable to what you want to do in your next job.

Job-Seeker's Glossary of Key Job-Hunting, Career, Job-Search, and Employment Terms

This glossary of job, career, and employment terms is designed to give job-seekers a quick definition.

Accomplishments — these are the achievements you have had in your career. These key points really help sell you to an employer — much more so than everyday job duties or responsibilities. In your cover letters, resumes, and job interviews, focus on key career accomplishments — especially ones that you can quantify.

Action Verbs — The building blocks of effective cover letters and resumes. These concrete, descriptive verbs express your skills, assets, experience, and accomplishments. Avoid nondescriptive verbs such as "do," "work," and forms of the verb "to be." Instead, begin each descriptive section with an action verb. Almost every resume book has a list of great action verbs to choose from.

Assessments — These tests ask you a series of questions and try to provide you with some sense of your personality and career interests. You shouldn't rely on the results of these tests by themselves, but the results can be a good starting point for discovering more about yourself and your interests and

Background Check — Used by employers to verify the accuracy of the information you provide on your resume or job application — and beyond. On the rise as prices fall on these services. Items checked include: employment verification, educational background/degrees, references, credit history, medical records, driving record, court records, criminal records, and more.

Benefits — An important part of your compensation package, and part of the salary negotiation process. Note that every employer offers a different mix of benefits. These benefits may include paid vacations, company holidays, personal days, sick leave, life insurance, medical insurance, retirement and pension plans, tuition assistance, child care, stock options, and more. Can be worth anywhere from 20 to 40 percent of your salary.

Career Activist — Someone who is proactive in planning, evaluating, directing, and controlling his or her career rather than simply reacting as situations arise. (Some call this approach career mapping.) A career activist has an enduring interest in understanding and achieving his or her full career potential, while maximizing career marketability.

Career Branding — Helps define who you are, how you are great, and why you should be sought out. Branding is your reputation. Branding is about building a name for yourself, showcasing what sets you apart from other job-seekers, and describing the added value you bring to an employer.

Career Change — Changing your occupation by devising a strategy to find new career choices. Most experts now predict that the average person will change careers three to five times over the course of his or her work life. Change may occur because you don't enjoy the work as much as you used to. Or maybe you can't progress further in your career.

Career Coach — Also called career consultant, career adviser, work-life coach, personal career trainer, and life management facilitator. These professionals have been likened to personal trainers for your life/career, serving the role as your champion, cheerleader, advocate, mentor, partner, and sounding board on all issues related to your job or career search.

Career Exploration — The process of finding a rewarding career path, as well as specific jobs within a particular career path. Think of career exploration and planning as building bridges from your current job/career to your next job/career.

Career Fair — There are many types of job and career fairs — from those scheduled during Spring Break for college students to industry-specific fairs for professionals — but they all have a common theme: a chance for a company to meet and screen a large volume of potential job candidates while simultaneously an opportunity for job-seekers to meet and screen a large number of employers.

Career Objective/Job Objective — An optional part of your resume, but something you should contemplate whether you place it on your resume or not. It can sharpen the focus of your resume and should be as specific as possible — and written in a way that shows how you can benefit the employer.

Career Planning — The continuous process of evaluating your current lifestyle, likes/dislikes, passions, skills, personality, dream job, and current job and career path and making corrections and improvements to better prepare for future steps in your career, as needed, or to make a career change.

Career Vision Statement — A set of career goals that a job-seeker sets for the long-term, typically five years or more. The purpose of a career vision statement is to give you a clear direction for the future; it is a vision that has been committed to paper to guide you in making future choices.

Cold Call — When a job-seeker approaches an employer (usually through an uninvited cover letter) who has not publicly announced any job openings

Compensation Package — The combination of salary and fringe benefits an employer provides to an employee. When evaluating competing job offers, a job-seeker should consider the total package and not just salary. See also Salary and Benefits.

Contract Employee — Where you work for one organization (and its salary and benefit structure) that sells your services to another company on a project or time basis. Compare to freelancer.

Corporate Culture — The collection of beliefs, expectations, and values shared by an organization's members and transmitted from one generation of employees to another. The culture sets norms (rules of conduct) that define acceptable behavior of employees of the organization. It's important for job-seekers to understand the culture of an organization before accepting a job.

Counter Offer/Counter Proposal — A salary negotiation technique used by job-seekers when a job offer is not at an acceptable level. Almost all elements of a job offer are negotiable, including the salary, non-salary compensation, moving expenses, benefits, and job-specific issues.

Cover Letter — Should always accompany your resume when you contact a potential employer. A good cover letter opens a window to your personality (and describes specific strengths and skills you offer the employer). It should entice the employer to read your resume.

Curriculum Vitae (CV) — Resume.

Declining Letter — A letter sent to an employer to turn down a job offer. The writer should keep the door open in case he or she would like to approach the employer again someday.

Degrees & Certifications — Recognition bestowed on students upon completion of a unified program of study, including high school, trade schools, colleges and

universities, and other agencies.

Diversity Job-Seekers — Numerous disadvantaged groups — women and minorities — often face extra challenges in the job-search.

Dress for Success — First coined by author John Malloy in the 1970s, the term Dress for Success signifies tailoring one's attire, grooming, and overall appearance toward making a great first impression in a job interview — as well as maintaining a professional look while on the job to aid career advancement. Will dressing properly get you the job? Not by itself, but it will give you a competitive edge and help you make a positive first impression

Electronic Resume (or E-Resume) — A resume (see resume) that is sent to the employer electronically, either via email, by submitting to Internet job boards, or residing on their on Web page. Includes numerous formats of resumes linked by their mode of delivery.

Elevator Speech — A a 15- to 30-second commercial that job-seekers use in a variety of situations (career fairs, networking events, job interviews, cold calling) that succinctly tells the person you are giving it to who you are, what makes you unique, and the benefits you can provide.

Email Cover Letter — A cover letter (see Cover Letter)

that is sent to the employer electronically via email. There are different rules that apply to writing these kind of cover letters, though the fundamental principles remain the same.

Employment Gaps — Are those periods of time between jobs when job-seekers are unemployed, either by choice or circumstances. Employers do not like seeing unexplained gaps on resumes, and there are numerous strategies for reducing the impact of these gaps on your future job-hunting.

Freelancer/Consultant/Independent Contractor — Where you work for yourself and bid for temporary jobs and projects with one or more employers. Freelancing is not an alternative to hard work, but many people enjoy the freedom, flexibility, and satisfaction of working for themselves.

Franchising — A legal and commercial relationship between the owner of a tradem ark, service mark, brand name, or advertising symbol (the franchisor) and an individual or organization (the franchisee) wishing to use that identification in a business.

Hidden Job Market — Only about 5-20 percent of all job openings are ever publicly known, which results in about four-fifths of the job market being "closed," meaning you can't find out about any new openings unless you do some digging. Strategies for uncovering the hidden job market include networking and cold calling.

Home-Based (Work-at-Home) Careers — Numerous opportunities exist for job-seekers who want more control over time and work, who want job flexibility to spend more time with family — by working from home. Unfortunately, this area is also one that has the most potential for scams and other fraudulent activities. R

Informational Interviewing — Just what it sounds like — interviewing designed to produce information. What kind of information? The information you need to choose or refine a career path, learn how to break in and find out if you have what it takes to succeed. It's the process of spending time with one of your network contacts in a highly focused conversation that provides you with key information you need to launch or boost your career.

Internships — One of the best types of work experiences for entry-level job-seekers because a majority of employers say experience is the most important factor in whether you're hired. Internships involve working in your expected career field, either during a semester or over the summer. Besides gaining valuable experience, you get exposed to the business environment and gain valuable references and network contacts.

Job Application — Sometimes also referred to as an Application for Employment. Many organizations require you to complete an application (either to get an interview or prior to an interview). Even though many of the questions duplicate information from your resume, it is extremely important to complete the application neatly, completely, and accurately.

Job Boards — Also referred to as Job Sites. There are five levels or types of job boards: general job boards (such as Monster.com and HotJobs.com), industry-specific job boards (such as TeachingJobs.com), geographic-specific job boards (such as AtlantaJobs.com), job-seeker specific "niche" boards (such as MBAJobs.com), and company career centers (such as HomeDepot.com)

Job Clubs — Sometimes known as networking clubs or job-finding clubs, enables you to expand your network of contacts — and also serves as a key support group when the job-hunt is longer or harder than you expected. A great tool for job-hunting, and job-seekers can either join an existing club or start your own!

Job-Hunting Etiquette — There are certain rules or protocols that should guide a job-seeker's conduct while job-hunting. Some people call these rules good manners, but more refer to them as business etiquette.

Job-Hunting on the Internet — Not a magic elixir that will guarantee that you find a job, but still a door to opportunities and techniques not available before the advent of the Net. Most job-seekers should spend no more than about 20 percent of their time and effort looking for a job online, though job-seekers in the technology/computer industry might be wise to spend up to 50 percent of their time looking for a job online.

Job Interviewing — All about making the best matches. Both the employer and the job-seeker want to determine if the fit is right between them. First impressions are key (see "dress for success"), and preparation is critical to interviewing success.

Screening interviews — usually conducted by a member of the human resources department, the screening interview is designed to weed out unqualified candidates. Providing facts about your skills is more important than establishing rapport.

Traditional interviews — uses broad-based questions such as, "why do you want to work for this company," and "tell me about your strengths and weaknesses." Interviewing success or failure is more often based on the job-seeker's ability to communicate and establish rapport than on the authenticity or content of their answers.

Behavioral interviews — based on the premise that past performance is the best indicator of future behavior and uses questions that probe specific past behaviors, such as "tell me about a time where you confronted an unexpected problem" and "tell me about an experience when you failed to achieve a goal."

Panel/group interviews — uses a committee of people, usually around a table, asking questions. The key to this type of interview is to balance eye contact with both the person who asked the question and the remainder of the group.

Case interviews — used primarily by management-consulting firms to determine how well suited you are to the consulting field. Case interviews measure problem-solving ability, tolerance for ambiguity, and communication skills along several dimensions. The idea is to find out how well you identify, structure, and think through problems

Situational interviews — sometimes also referred to as a scenario-based (problem-solving) interview, where the job-seeker is placed in a hypothetical situation (such as dealing with an irate customer), and is judged by how well s/he reacts to complex information and ability to resolve problem and arrive at solutions.

Stress interviews — usually are a deliberate attempt to see how you handle yourself under pressure. The interviewer may be sarcastic or argumentative, or may keep you waiting. Expect these things to happen, and when it does, don't take it personally. Calmly answer each question as it comes. Also called intimidation interviews.

Phone interviews — have only one purpose: to decide if there is a good enough match to justify a site visit. Make sure to set a specific time for your telephone interview — not just "sometime this week."

Job Satisfaction — A term to describe how content an individual is with his or her job. It includes many factors, including the work itself, value to the organization, impact on organization, compensation, and more.When workers are very unhappy with their jobs, they suffer both mentally and physically.

Job Search Agent — A program offered by many job boards that allows job-seekers to passively search for jobs by selecting criteria for new job postings. At some time interval, the program emails the job-seeker a list of new job postings that fit the criteria, allowing the job-seeker to decide whether to take any action.

Job-Search Domino Effect — States that five key phases comprise any good job search, and if you ignore any one of them or conduct one poorly, the likelihood of a successful job search decreases dramatically — just as if you pulled a domino out of a row of dominos.

Job Shadowing — One of the most popular work-based learning activities because it provides job-seekers with opportunities to gather information on a wide variety of career possibilities before deciding where they want to focus their attention. Job shadows involve brief visits to a variety of workplaces, during which time you "shadow," observe, and ask questions of individual workers.

Job Skills — The skills you need to do a particular job. For example, an accountant needs to have good math and accounting skills; a doctor needs to have good medical, scientific, and personal skills.

Job Skills Portfolio — Also referred to as a Career Portfolio, a job-hunting tool a job-seeker develops to give employers a complete picture of who you are, including samples of your work — your experience, your education, your accomplishments, your skill sets — and what you have the potential to become — much more than just a cover letter and resume can provide.

Key Accomplishments — An optional part of your resume, but one that is growing in use — especially with scannable (text-based) resumes. This section should summarize (using nouns as keywords and descriptors) your major career accomplishments. Sometimes also referred to as "Summary of Accomplishments," "Qualifications Summary," or simply "Accomplishments."

Keywords — Nouns and noun phrases that relate to the skills and experience that employers use to recall resumes scanned into a database. Keywords can be precise "hard" skills — job-specific/profession-specific/industry-specific skills, technological terms and descriptions of technical expertise, job titles, certifications, names of products and services, industry buzzwords, etc

Letter of Acceptance — Used to confirm the offer of employment and the conditions of the offer; i.e., salary, benefits, starting employment date, etc. It is always a good idea to get the entire offer in writing.

Letter of Agreement — A brief letter outlining the conditions of employment. Whether initiated by the employer or the candidate, it is always a good idea to get your entire offer in writing. Sometimes is form-based or may even be an employment contract. See also salary and salary negotiation.

Letter of Recommendation — A letter of support for your skills, ability, and work ethic, usually written by a former boss or co-worker, but could also be from a teacher or personal reference. Good for applying to graduate school, but seen as fairly worthless in job-hunting because no one who would write you a recommendation letter would say anything negative about you. See reference list.

Low-Wage Worker/Low-Wage Job — Call them low-wage workers employed in low-paying, minimal or no-benefit jobs. Call them exploited or the working poor. Call them living (barely) from paycheck to paycheck with no job security. Call them under-educated, under-trained, and under-respected. Call them job-seekers stuck in dead-end jobs. But, whatever you do, do not call them low-skilled or lazy.

Mentor — A person at a higher level within a company or within your profession who counsels you and helps guide your career. Some organizations have formal mentoring systems, while most informal mentoring relationships develop over time. A mentor relationship is one where the outcome of the relationship is expected to benefit all parties in the relationship for personal growth, career development, lifestyle enhancement, spiritual fulfillment, goal achievement, and other areas mutually designated by the mentor and partner.

Moonlighting — The experience of working multiple jobs (also referred to as dual or multiple jobholding). People working multiple jobs come from just about every demographic group. Appears to be on the rise.

Networking — Involves developing a broad list of contacts — people you've met through various social, professional, and business functions — and encouraging them assist you in looking for a job. People in your network may be able to give you job leads, offer you advice and information about a particular company or industry, and introduce you to others so that you can expand your network.

Non-Traditional Careers — Careers in which fewer than 25 percent of the workforce is of one gender. There are many pros and cons for job-seekers considering working in a non-traditional career path.

Offer of Employment — An offer by an employer to a prospective employee that usually specifies the terms of an employment arrangement, including starting date, salary, benefits, working conditions. Also called a job offer.

Older (Mature) Workers — Job-seekers who are older than 45 face more challenges in the job-search than other types of job-seekers.

Overqualified — A label employers often use on mid-career job-seekers who appear to have one of three flaws: too many years of experience, too much education, too highly paid in current or previous job.

Passive Job-Search — A strategy where employed workers stay prepared for new job and career opportunities by maintaining a current resume, continuing to network, staying registered with one or more job-search agents. You are not openly on the job market, but keep an interest in new possibilities.

Personal Mission Statement — Helps job-seekers identify their core values and beliefs. Writing a personal mission statement offers the opportunity to establish what's important and perhaps make a decision to stick to it before we even start a career. Or it enables us to chart a new course when we're at a career crossroads.

Portfolio Career — A situation where instead of working a traditional full-time job, job-seekers work multiple part-time jobs (including part-time employment, temporary jobs, freelancing, and self-employment) with different employers that when combined are the equivalent of a full-time position. Portfolio careers offer more flexibility, variety, and freedom, but also require organizational skills as well as risk tolerance.

Quarterlife Crisis — A period in your twenties marked by high anxiety about your career — and finding your "true" career, multiple job and/or career changes, fears and self-doubt about achieving career and personal goals, depression, and feeling lost or adrift. Experts say that the crisis hits folks in their twenties, because after years of learning the system of how to succeed in school, college grads are thrown into the world of work with no real understanding of how to succeed in it.

Recruiters/Headhunters/Executive Search Firms — Professionals who are paid by employers to find candidates for specific positions. They often recruit candidates, but job-seekers can also approach them. Often specialize by industry or geographic region. Avoid any firms that require you to pay for their services.

Reference List — Sometimes also referred to as a Reference Sheet. Simply a listing — with key contact information — of your references. Never include references on your resume or cover letter; they should be listed on a separate references sheet that matches the look of your resume. Never provide a list of references to an employer unless you are requested to do so

References — A group of people who will say good things about you and who know specifics strengths that you offer. Can include work references (current and past supervisors), educational references (former teachers or school administrators), and personal references (who can speak of your character). Always ask people before including them as a reference for you.

Researching Companies — The process of gathering information about a company, its products, its locations, its corporate culture, its financial successes. This information is extremely valuable in a job interview where you can show off your knowledge of the company, and can also help you in writing your cover letter.

Resigning/Resignations — When you decide it's time to quit your job (also referred to as giving notice), it's always better to submit your official resignation — with your industry's customary amount of notice. Whenever possible, do not leave on bad terms with your employer.

Resume — A key job-hunting tool used to get an interview, it summarizes your accomplishments, your education, as well as your work experience, and should reflect your special mix of skills and strengths.

Curriculum Vitae — also called a CV or vita and similar to a resume, but more formal, and includes a detailed listing of items beyond the typical resume items, such as publications, presentations, professional activities, honors, and additional information. Tends to be used by international job-seekers, and those seeking a faculty, research, clinical, or scientific position.

Salary — Financial compensation an employee receives for performing the job, and part of your compensation package. Can be determined by hourly, daily, weekly, monthly, and yearly. Also can include overtime pay, bonuses, and commissions. See also Benefits and Compensation Package.

Salary History — Some employers will request that you submit a salary history. A salary history tells them the level and frequency of your promotions. It should be separate page from your resume or cover letter. Be sure to include the full compensation you received in each job, not just salary information. By providing a salary history, you sometimes place yourself in a precarious position of either pricing yourself out of the position or appearing to be at a lower level than the company seeks.

Salary Negotiation — An extremely important process in which job-seekers attempt to obtain the best compensation package possible, based on skills and experience, the industry salary range, and the company's guidelines.

Salary Requirements — Some employers may ask you to state the salary you require for a specific job opening. You've got to be careful here. If your salary requirement is too high, you won't get an offer. If it's too low, you won't get what you're worth. The best strategy is to state that you're open to any fair offer and are willing to negotiate.

Summer Jobs/Part-Time Jobs for Teens — Whether it's to gain experience, earn some spending money, or save for college, getting a summer or part-time job is almost a rite of passage for most teens. Teens are often limited to certain types of jobs and hours worked per week.

Survival Job — Typically a low-end, low-paying job that a displaced job-seeker takes on a temporary basis (often as a last resort) to cover basic living costs, in order to survive and avoid bankruptcy — or worse.

Telecommuting — Also referred to as Teleworking, is a employment arrangement where the employee works one or more days from a remote location, often an office in the employee's home. For job-seekers seeking increased job flexibility and reduced commuting times and costs and for employers seeking a better balance of morale and work efficiency.

Temping — Working short employment stints with a variety of clients, usually through a temping agency or staffing firm. Previously temps were mostly administrative, but job-seekers can now find temping agencies covering most professions. Temping is great for building resume, learning skills, networking — and job flexibility and variety. See also Temporary Agency.

Temporary (Temp) Agency/Staffing Firms — Companies that place workers in jobs on a contract or temporary basis. Some provide training. Many are specialized (professional, clerical, computing, accounting, etc.).

Testing — An increasing number of employers are using a variety of career and skill-based tests to screen job applicants. Thus, you may be asked to take any number of tests during your job search, from aptitude and personality tests to honesty and drug tests.

Thank You Letters — After every interview, you should send a letter thanking each person who interviewed you. It's just common courtesy, and only a small percentage of job-seekers actually perform this crucial ritual, so you'll stand out from the crowd.

Transferable Skills — Skills you have acquired during any activity in your life — jobs, classes, projects, parenting, hobbies, sports, virtually anything — that are transferable and to what you want to do in your next job.

Underqualified — The underqualified or just plain unqualified label most often plagues new graduates with limited experience, as well as career-changers whose experience is outside the area they now wish to pursue.

USP — An advertising term — unique selling proposition — that refers to the one thing about a product that makes it distinct from all others. In job-hunting, job-seekers need to find the one thing that makes you more qualified for this job than anyone else. What can you offer that no other applicant can?

Workaholism — a condition where work becomes all-consuming, becoming more important than personal or family obligations. The person lives for his or her work, has only work friends, rarely takes time off, and is always in communication with the office. Does not function well outside work.

Workplace Values — Concepts and ideas that define a job-seeker and influence your satisfaction — not only with your job, but with your life. Job-seekers should perform a values check every few years to make sure your career is on track.

10 Steps to a Successful Interview

- Arrive on time.
- Introduce yourself in a courteous manner.
- Read company materials while you wait.
- Have a firm handshake.
- Listen.
- Use body language to show interest.
- Smile, nod, give nonverbal feedback to the interviewer.
- Ask about the next step in the process.
- Thank the interviewer.
- Write a thank-you letter to anyone you have spoken to.

What Employers Want?

Employers say they are impressed by job candidates who have excellent communication skills, good grooming habits, and relevant work experience. Employers say they want trustworthy new hires who can move right in, get along with their co-workers, and get the job done without having to be babied at each step.

Top 10 Qualities Employers Seek	Employers Rate the Importance of Experience
Communication skills (verbal and written) Honesty/integrity Teamwork skills (works well with others) Interpersonal skills (relates well to others) Motivation/initiative Strong work ethic Analytical skills Flexibility/adaptability Computer skills Organizational skills	Relevant work experience ....4.0 Internship experience .... 3.9 Any work experience ....3.6 Co-op experience ....3.2 (5-point scale: 5=Extremely important; 1=Not important)

Interviewers' Favorite Questions...and Answers

You're wearing your best interview suit and facing your best friend, who's wearing the most inscrutable hiring-manager face she can muster. You've carefully positioned a video camera to record your every move. All is in place for your mock interview.

"Tell me about yourself," your friend/interviewer intones, adjusting her glasses and gazing steadily into your eyes.

What should you tell her? What would you tell a real recruiter or hiring manager?

"Don't tell me where you were born and raised," says a manager and a veteran of countless mock interviews with students. "Don't tell me that you were a house captain. Focus on your academics and experience. Ask yourself, 'what are the top five things I want this person to know about me?'"

He says that while many recruiters ask questions that are a bit more pointed than "tell me about yourself," it's still likely to come up in many interviews and it's best for students to prepare for it.

What other kinds of questions do recruiters ask? Following are 10 more, plus ideas for how to answer or the kinds of competencies the interviewer is seeking.

1. **What do you see yourself doing five years from now?**

   "I want to hear something related to the job,"says a relations manager for a call center. "I don't want to hear 'I want to be an astronaut' or 'I want to win the Filmfare Award.'"

   People say the question is designed to help the interviewer know if the job seeker will be happy in that position, or if he or she wants to work in it only as long as it takes to find something "better."

2. **How do you make yourself indispensable to a company?**

   "We are looking for both technical and interpersonal competence," says a recruiting and advertising manager

   She explains that students who have interned or completed cooperative education assignments generally answer the question best because they know what working for a company entails.

3. **What's your greatest strength?**

   Don't just talk about your strength—relate it to the position. Let them know you are a qualified candidate.

4. **What's your greatest weakness?**

   Say something along the lines of, "I have difficulty with this thing, and these are the strategies I use to get around it".

For example, you could say, 'I'm not the most organized of individuals, so I always answer my e-mails and phone calls right away. I'm aware of the problem and I have strategies to deal with it."

5. **Tell me about a time when your course load was heavy. How did you complete all your work?**

   This involves getting the interviewee to describe a situation that includes a task that needed to be accomplished, the action taken to accomplish the task, and the result of that action.

6. **Tell me about a time when you had to accomplish a task with someone who was particularly difficult to get along with.**

   They want to hear something that shows the candidate has the ability to be sensitive to the needs of others but can still influence them "Don't say 'I just avoided them' or 'They made me cry.'"

7. **How do you accept direction and, at the same time, maintain a critical stance regarding your ideas and values?**

   Internship or co-op experience can give students the experience to answer that question, pointing out that students with good interpersonal skills honed on the job can understand how to walk that fine line

8. **What are some examples of activities and surroundings that motivate you?**

   Give examples of things around you which heave actually motivated you.

9. **Tell me how you handled an ethical dilemma.**

   The best way to handle such a situation would be to go to a supervisor, explain the situation, and ask for advice. He adds that students who can't offer a situation that they handled correctly the first time can explain how they learned from making mistakes.

10. **Tell me about a time when you had to resolve a problem with no rules or guidelines in place.**

    "I'm looking for a sense of urgency in initiating action," says a manager, explaining that the question probes a student's ability to overcome obstacles.

    "I'm looking for the right thing in terms of customer service," he says.

Dress For Success

Its true - You Are What You Wear! Especially when it comes to the job interview.

Believe it or not, you are examined from head to toe in all interviews. Not only is your experience and personality sized up...but also your appearance. The dress code for men and women differ substantially.

Before you go on that job interview, make sure you read these dressing tips! Many people take for granted the appearance aspect of the job hunting process. The better you are dressed, the more confident you will feel and exude.

There's no impression like the first impression. Initial decisions made about you in the first three minutes of the job interview are nearly irreversible.

The purpose of your clothing is to project an image of professionalism.

Consider

Your clothes should subtly add to your appearance. More importantly, clothing shouldn't detract from your business persona.

Much as you may want to stand out, wearing the business

"uniform" communicates instantly that you are a mature, stable professional and a member of the team.

Even if the internal dress code is very casual, your interviewing attire should adhere to a conservative standard.

Dress up - the best you would ever dress when actually employed there.

Here's how to dress for the best possible outcome:

Research the prospective employer - companies and even industries have definable corporate cultures. Find out what the corporate uniform is - and wear it.

- Select an outfit you have worn before and are comfortable wearing.
- Try on the outfit a few days before - enough time to have drying-cleaning and repairs completed.
- Check the ensemble for missing buttons, frayed cuffs and other needed repairs.
- Everything must be clean, neat and ironed.
- No bulging pockets or sagging coat lining.
- Hair and nails must be clean and groomed.
- Scent should be low-key or absent.
- The favorite color of most interviewers is blue - it conveys trust, calm and confidence. Blue is a very good interview color.

The Fine Points

Shoes should be comfortable and polished. Shoes that are well cared-for signal "Good attention to detail."

Be parsimonious with scent - many people are allergic and too much perfume is an instant interview killer.

Jewelry - real jewelry, if you have it. No jewelry is better than fake. No rhinestones cufflinks, no cloth watchbands, no novelty tie tacks.

Clean and polish your briefcase or purse; organize the inside.

A winter coat must be cleaned and pressed, particularly since a coat may be the main item your interviewer first sees.

Your umbrella should be in conservative colors (black, tan, navy, gray) and in working order. Do not leave it to drip on the company carpet.

Bring a pen and paper; check that the pen works and doesn't leak. Store them in the inside jacket pocket, where you can easily find them.

Allow time to dress with care and deal with emergencies. Examine the results, front and back, in a mirror.

Men - Consider These Tips/Advice

- tailored suits in navy, gray, beige. (black is a funeral color, avoid it.)
- dark suit, light shirt
- natural fabrics - wool/wool blend for the suit, cotton for the shirt, silk for the tie
- business shoes and over -the-calf dark socks
- matching silk tie in low-key colors

Women - Consider These Tips/Advice

- simple, tailored suit
- tailored dress
- dress & jacket combination
- simple blouse
- natural fabrics - wool/wool blend, cotton and silk
- conservative colors - blue, gray, beige or black
- avoid loud or flashy styles and colors
- use makeup sparingly.

- low -heeled pumps, flesh colored stockings

How To Master Telephone Interviews

Don't be afraid to pick up the phone! The telephone interview is by far becoming more and more popular. Many job hunters still get that adrenalin rush even with the phone interviews.

But following the tips and advice in this article will help you master the phone interview and get you to the next step - the face to face interview.

Often, the first step in the hiring process is the telephone interview.

Companies and the recruiters they employ use the telephone interview to develop a pool of candidates to look at closer, and to pare down the number of applicants for a job opening.

The advantages to the company are:

- the cost is less.
- the list of questions can be standardized.
- the interview can be delegated to a lower level (cheaper) employee.
- it can be done quickly.

The aims on both sides of the telephone are limited. The caller wants a selection of qualified candidates, and the process screens out many candidates. If the call is a straightforward screening call, the caller will likely ask about your experience, availability and salary requirements. Your strategy is to provide facts that support your resume, with some context about your performance. Try using numbers and facts to be effective, however, you don't want to volunteer anything that could disqualify you . Make every effort to sound professional but not personal, as this call is not to establish rapport.

Since you are unlikely to win the job from a telephone interview, your goal is to secure an in-person interview with the person who has the authority to hire. Approach the call with that attitude.

Managing the Telephone Interview

Try to reschedule surprise interviews. Say that you have a conflict and suggest a time you can call back. When you call back, be prepared for the call just as you would for a full-dress interview.

Have ready

pen and paper, a calculator

the job ad and the resume and cover letter which you sent in response to the ad.

a list of your accomplishments which relate to the job you are discussing.

research you have done on the company.

a short list of questions about the job.

your calendar.

The Techniques of a Pro

Smile - it comes through in your voice.

Speak directly into the phone.

Don't smoke, chew gum, eat or drink anything. It all telegraphs to your listener.

Stand up. Your voice sounds stronger.

Avoid ah, er, hum. This habit is especially noticeable on the telephone. This takes practice. So practice.

For a winning performance

Confirm the caller's name and company. Get the caller's telephone number.

Be aware that the caller can't see you - can't see your hand gestures, can't see you taking notes.

Pace the call. Let the caller do most of the talking, without interruptions.

Do use the technique of repeating or re-phrasing questions. It tells the caller that you listened carefully, and gives you time to think about your answer.

Avoid the simple yes or no; add selling points at every opportunity.

If you need time to think, say so - as in radio, silence during a telephone conversation is dead air time.

Compensation issues come at the end of the interviewing cycle, never at the telephone stage. You can truthfully say you don't know enough about the job to state a salary figure. And, of course, you would need a personal interview to really talk with the company. Which is another way to go for the personal interview. Re-affirm your qualifications, express your interest in the job and the company. Say you would appreciate the opportunity to talk about the job further - in person.

How To Answer The TOUGHEST Interview Questions!

What the Interview is After?

Few interview questions invoke as much dread as "Tell me about your greatest weakness?" Keep in mind that by asking this question that the interviewer is trying to determine if you engage in self analysis and how you handle difficult questions.

These 5 quick tips will help you to successfully navigate this interview hurdle on your way to landing your ideal job.

5 Tips To Help Answer - "TELL ME ABOUT YOUR GREATEST WEAKNESS?

Tip #1 - Mention The Downside Of A Skill.

This strategy is successful because you actually answer the weakness question by showcasing a skill. For example "I am very committed to customer service. Sometimes this leads to my spending quite a bit of time with customers. At times this causes me to fall behind in other areas."

Tip #2 - Share A Weakness That You Are Working To Overcome.

This tactic is successful because it shows that you are committed to self-improvement. An example of this is "I am not really comfortable presenting before an audience. I have joined a public speaking group in order to overcome some of my nervousness." Another example is "I currently only speak English. I recognize that the world is getting smaller day by day and I am studying Spanish in order to be better prepared for the future."

Tip #3 - Use A Weakness That Has Nothing To Do With The Job.

While it is usually preferable to answer with the downside of a skill or a weakness you are working on, sometimes, the best tactic might be to use a weakness that is unrelated to the job. This way you can honestly answer the question while not casting doubt on your ability to fulfill the job requirements. Examples include "I am a terrible cook" or "I cannot dance." This strategy is useful if you are trying to inject a (quick) humorous story into the interview regarding some mishap related to the weakness.

Tip #4 - Do Not Pretend To Be Perfect.

It is never a good idea to answer this question by saying you have no weaknesses. Everyone has weaknesses. By saying you have none, you will appear arrogant.

Tip #5 - Do Not Give A Canned Or Common Answer.

This will make you will seem phony and insincere at worst and unimaginative at best. An example of a very over-used answer is "I am a perfectionist." Say this or something similar and be prepared for the interviewer to (inwardly) roll his/her eyes and subtract points for originality.

When talking about your weaknesses, always clearly communicate both what you are doing to overcome the weakness and that you are confident that your efforts will be successful.

No, it's not time to throw your resume in the trash and start a "new age job search". But one thing that any job seeker must understand is that the showcase of talents does not begin and end with the resume. There are many "secret" abstract, often called "soft", skills that employers keep an eye out for.

The five key "secret skills" that interviewers examine and how to demonstrate them in an interview situation.

These five skills are:

1. Organizational
2. Critical Thinking
3. Communication
4. Interpersonal
5. Multi-Tasking

1. Organizational Skills

Unless you are applying for a job as a mad scientist, organization is an essential skill for any job. Employers can get sense of how an individual will handle large workloads by how organized that person is during the interview. Moreover, a person that makes a sincere effort to stay organized is an employee that will take a job seriously and make a sincere effort to get things done.

The best way to display these skills:

Dress professionally and neatly for an interview.

Keep supplies or materials on hand if you think they might be pertinent to the interview. This can go beyond pen, paper, resumes, and business cards depending on the position you apply for.

Organize your thoughts before the interview. Preparation for typical interview questions will reflect a sense of general readiness.

2. **Critical Thinking Skills**

Nobody wants a mindless drone for an employee. If they did, they would buy a robot. Employers want people that can think on their feet and respond. They are looking for people that won't come crying with every little setback. They are looking for problem solvers. Having critical thinking skills means that you can come through in the clutch.

The best way to display these skills:

Prior to the interview, prepare of a list of anecdotes or previous jobs that required critical thinking to solve a problem. When applicable, bring these examples up in the interview.

Talk your way through the answers. Let the interviewer understand your train of thought when responding to questions. This can also buy you a little extra time if you are unsure of how to answer.

3. **Communication Skills**

Glossophobia, or the fear of public speaking, is the number one fear but making an impact requires these skills. Unless you can communicate ideas to others effectively, you may not come across as very confident. This is precisely why so many employers ask for individuals with good communication skills, often including public speaking.

The best way to display these skills:

Practice speaking, or answering interview questions in a mirror. This will get you used to speaking aloud and let you see the things you may be doing wrong.

Practice interviews with another person, so you can learn to keep cool when reacting to another person's comments.

Stay calm and ALWAYS MAINTAIN EYE CONTACT. It's

hard to disagree with a confident person. Once you SEEM confident, you hold all the cards.

4. Interpersonal Skills

Along with being able to communicate your own ideas well, you have to be receptive to other ideas and work constructively with them. Companies need versatile team players: people that will work hard on their own and increase the depth and effectiveness of a group effort.

The best way to display these skills:

As in the case of critical thinking, it is a good idea to prepare a list of examples in which you were part of a successful team effort. These items may not be on your resume, but could come up in an interview.

When possible, reflect back on cases where you coordinated a team effort. It is one thing to work well in a group, but it is even better when you show that you can also lead and take charge of a group.

Don't be afraid to mention troubles within a team that you had to overcome. A group of people will not agree on everything 100% of the time. Being able to work through problems and succeed is paramount.

5. MULTI-TASKING SKILLS

Businesses are always happy to drive down costs, and the best way to do this is by hiring fewer individuals who can multi-task. It is often the case that one efficient employee can do the work of two typical employees. Employees are paid for the hours they work, and employers want to get the most out of what they pay. An employee that can complete multiple tasks at once is the solution.

The best way to display these skills:

When discussing previous positions held, include situations where you worked on multiple tasks at the same time.

Prepare a list of projects that required you to separate tasks into clusters that could be addressed simultaneously. Be ready to explain the thinking behind your separation system.

Show a willingness to take on many responsibilities. Any worker can pick up one or two, but if you can pick up more without getting spread to thin, you become a valuable asset.

Conclusion

The resume will always be around and serve as your primary means of communicating skills with a prospective employer. But remember that you are more than just a list of skills on a piece of paper. The interview lets the employer see whats not easily determined from a resume and also your chance to shine. Mastering the art of showcasing your "secret skills" will let an interviewer know you are person they need to hire.

Avoiding Potential Interview Disasters

There are definitely things that you can do to avoid minor mishaps which could ultimately blow an interview. Become familiar with these 7 potential interview disasters so you can prevent them from obstructing your path to that ideal job.

1. Don't Arrive-late

Showing up late is both rude and inconsiderate. Is this the first impression that you want to leave with a potential employer? Map out your route and try it out before the interview. Plan on being at least half an hour early to your appointment. This will provide a buffer to protect against wrong turns, traffic jams and all the other mishaps that may befall you. If you arrive early, you can use the time to calm your nerves.

2. Don't Say The Wrong Name

Many a nervous candidate has been known to accidentally call the interviewer the wrong name. In order to avoid this disconcerting faux pau, find out who you will be speaking to before the interview. Memorize the name(s). If this information is not available prior to the meeting, then write the person's name on your notepad as soon as you sit down for the interview. If you do slip-up, do not make a huge fuss. Apologize quickly (and sincerely) and move on.

3. Don't Say the Wrong Thing

Choose your words carefully. Avoid impulsive answers; the first thing that pops into your head may not be the best response. Remember, it's ok to pause if you need some time to think. Feel free to say "that's a good question; let me take a moment to think about it." This demonstrates that you think before you speak. Is your everyday speech peppered with expletives or other potentially offensive phrases? If so, take care to avoid these during your meeting.

4. Don't Become Speechless

Interviews are stressful situations for even the most qualified candidates. This tension can lead to candidates "freezing up" during the meeting. Alleviate some of the expected stress by practicing mock interview questions. Have a friend conduct simulated interviews. If possible, have him/her conduct the interview in a variety of manners including reserved, rushed, and disinterested. This way you will be better prepared for whatever the interview may bring.

5. Monitor your Internal System

Nature has a funny way of acting up at the wrong moments. Fortunately, you can help prevent these unwanted incidents. Avoid the awkwardness of a growling stomach by eating a few hours before the interview. Be careful in what you eat and drink in the 24 hours prior to the interview. Do not overindulge; an upset stomach or hangover is formidable distraction.

6. Don't Be Too Modest

The interview is no place for humbleness. Too much modesty can make you appear introverted or lacking confidence. Don't be afraid to be your own cheerleader. Prior to the interview, make a list of your accomplishments both personally and professionally; practice talking about them. Have a friend listen to your answers as you practice. This will help prevent you from crossing the line between justifiable pride and boasting.

7. Don't Lack Enthusiasm

No one wants to work with a stick in the mud. With this in mind, how can you prevent from appearing lukewarm? Smile and maintain eye contact. Sit forward in your chair. Avoid speaking in a monotone. Be positive in your responses.

Conclusion

By preparing yourself against these potential interview disasters you are one step closer to getting the job of your dreams. Remember sometimes it really is the little things that make the difference.

Interview Preparation Guide

Recruitment Process

The recruitment process at a call center requires a potential candidate to go through a series of rounds involving telephonic interview, online tests to assess technical & language skills, V&A assessment and HR screening.

A candidate typically needs to go through the following screening rounds:

1st Screening	:	Telephonic Interview
2nd Screening	:	Online Tests
3rd Screening	:	Technical Interview

Final HR Screening

Interview Checklist

Given below is a checklist to ensure that your interview at is a good experience:

- Do your homework:

- Checkout the website to learn more about the organization
- Ask yourself why you are a good fit for the job you are applying for
- Review your resume thoroughly and be prepared to discuss it with the interviewer
- Do a prior-check on the interview location so that you are on-time for the interview
- Dress smartly and come stress free and ready to share your experience with us
- Keep your resume short & crisp (between one to three pages) and include relevant experience and educational background, including lengths of service (with dates) and accomplishments
- Remember to carry two copies of your latest resume when you come for the interview

References:

Use two-three former supervisor/ peer references. Academic references are acceptable if you do not have recent work experience

Please request approval from references prior to submission and cross-check on latest telephone numbers & e-mail addresses

Other Items:

One copy each of your relevant educational transcripts (graduation mark sheet or degree)

- Two copies of your latest photograph
- Your latest salary slip (if already employed)
- During the Interview:
- Switch off cell phone before the interview begins
- Be prepared to discuss specific examples demonstrating

your skills, experience and qualifications

- Feel free to ask your interviewer questions
- Ask for a feedback date

Introduction Round-Round 1

This round includes speaking on a topic for one minute. This round is usually after the introduction round and is the basis of selection of the candidate.

The following are evaluated: -

Fluency

Content

Grammar

Accent

Interview Question, "Tell Me Something About Yourself?"

It's a question that most interviewees expect and it is the most difficult to answer as well. Though one could answer this open- ended question in a myriad of ways, the key to answering this question or any other interview question is to offer a response that supports your career objective.

This means that you shouldn't respond with comments about your hobbies, spouse, or extra curricular activities. Trust me, interviewers aren't interested. To start with there is no correct answer to this interview question. I would lean in the favor of a quick reference to some personal traits that give a quick-view of who you are. From there one could move to a one sentence of any relevant education/qualification. There should also be a mention of employment history.

Purpose of the Question:

In one of my write-up "Across the interview table" I did mentioned that there is a purpose of asking each and every question in the interview. One cannot ask anything and everything. Again, set of interview questions varies from industry to industry and position to position. Interviewers use the interview process as a vehicle to eliminate your candidacy.

Every question they ask is used to differentiate your skills, experience, and personality with that of other candidates. They want to determine if what you have to offer will mesh with the organization's mission and goals.

What type of answer is Expected???

Try to avoid this type of answer: I am a hard-worker who is good with numbers. After I worked as a financial analyst for a few years, I decided to go to law school. I just finished and now am looking for a new challenge.

Speak something like this:

I began developing skills relevant to financial planning when I worked as a financial analyst for three years. In that role, I succeeded in multiplying the wealth of my clients by carefully analyzing the market for trends. The return on the portfolios I managed was generally 2% more than most of the portfolios managed by my company. My initiative, planning, and analytic skills were rewarded by two promotions. As the manager of a team, I successfully led them to develop a more efficient and profitable strategy for dealing with new accounts. My subsequent training in the law, including tax law and estate law, gives me an informed view of what types of investments and charitable gifts would be most advantageous for your clients.

Preparing for the Answer:

Follow the following steps as outlined below to ensure your response will grab the interviewer attention.

1. Provide a brief introduction. Introduce attributes that are key to the open position.
2. Provide a career summary of your most recent work history. Your career summary is the "meat" of your response,

so it must support your job objective and it must be compelling. Keep your response limited to your current experience. Don't go back more than 10 years.

3. Tie your response to the needs of the hiring organization. Don't assume that the interviewer will be able to connect all the dots. It is your job as the interviewee to make sure the interviewer understands how your experiences are transferable to the position they are seeking to fill.
4. Ask an insightful question. By asking a question you gain control of the interview. Don't ask a question for the sake of asking. Be sure that the question will engage the interviewer in a conversation. Doing so will alleviate the stress you may feel to perform.

There you have it - a response that meets the needs of the interviewer AND supports your agenda.

When broken down into manageable pieces, the question, "So, tell me about yourself?" isn't overwhelming. In fact, answering the question effectively gives you the opportunity to talk about your strengths, achievements, and qualifications for the position. So take this golden opportunity and run with it!

When Asked by Different People?

HR manager or CEO of the company or the Departmental Head can ask the same question and your answer should vary. The expectation of each such person is different.

When asked by HR Manager your response must be like this: "My career has been characterized by my ability to work well with diverse teams. I seek out opportunities to involve others in the decision-making process. This collaboration and communication is what has enabled me to achieve success in my department. People are the most valuable resource of any organization."

When asked by CEO your response must be like this: "I

have achieved success in my career because I have been focused on the bottom line. I have always sought out innovative solutions to challenging problems to maximize profitability. Regardless of the task or challenge, I always established benchmarks of performance and standards of excellence. I have never sought to maintain the "status quo." An organization that does not change and grow will die. I would enjoy working with you to help define new market opportunities in order to achieve the organization's goals."

3
VOICE AND ACCENT FOR CALL CENTERS

There are mainly two types of accents for call centers: -

- American Accent
- British Accent

To be sure that your accent is acceptable in a call center, you need to work on your accent and neutralize it.

Differences Between American and British English

American English is the form of English used by people in the United States and, as a lingua franca or second language, by people in many parts of the world.

American English does not include Canadian English; although Canadian pronunciation and vocabulary is very similar to that of the United States, Canadian spelling often takes the Commonwealth form. Regions and countries that tend to use American English in teaching and publishing include much of Eastern Europe (including Russia), the Far East (especially Japan, South Korea, Taiwan, and the Philippines; although largely excluding the former British colonies of Hong Kong and Singapore), the Americas (excluding other former British colonies such as Canada, Jamaica, and the Bahamas) and, in Africa, Liberia, and Namibia. The World Bank, and the Organization of

American States, among other organizations, also use American English.

Commonwealth English is the form of English used across the United Kingdom and most of its former colonies in the British Empire, most notably in much of Africa (including South Africa and Egypt), the Indian subcontinent (Pakistan, India, and Bangladesh), Malta, Australia and New Zealand, and portions of Southeast Asia (Myanmar, Singapore, Malaysia and Thailand), as well as Hong Kong. Like American English, it is used in many parts of the world as a lingua franca. This form of English, mostly interchangeable with British English, is sometimes called Commonwealth English because most of its speakers live in nations that are members of the Commonwealth of Nations. Other nations and groups that use British English include the European Union and, often, the United Nations. Many international organizations, like the International Olympic Committee, NATO, the World Trade Organization and ISO also use British English.

The forms of English spoken in Canada exhibit features of both British and American dialects; while spelling is closer to British English, pronunciation and vocabulary are much closer to American English. Many words and phrases thought of as "Americanisms" are also used by Canadians.

Although British English is a term used when describing formal written English used in the United Kingdom, the forms of spoken English used in the United Kingdom vary considerably more than in most other areas of the world in which English is spoken, even more than in the United States, despite the vastly larger population and land area of the U.S. Dialects and accents vary not only within regions of the UK—for example, in Scotland, Northern Ireland, and Wales—but also within England, with differences occurring within the space of a few miles in some cases. The written form of the language universally taught in schools is Commonwealth English with a slight emphasis on a few words which might be more common in the specific areas than others. For example, although the words "wee" and "small" are interchangable, one is more likely to see "wee" written by a Scot than by a Londoner.

Although spoken American and British English are generally mutually intelligible, there are enough differences to occasionally cause awkward misunderstandings or even a complete failure to communicate. George Bernard Shaw once said that the United States and United Kingdom are "two countries divided by a common language"; a similar comment is ascribed to Winston Churchill. Henry Sweet predicted in 1877 that within a century, American English, Australian English and British English would be mutually unintelligible, but it may be the case that increased world-wide communication through radio, television, the Internet, and globalization has reduced the tendency to regional variation. This can result either in some variations becoming extinct (as, for instance, truck has been gradually replacing lorry in much of the world) or in the acceptance of wide variations as "perfectly good English" everywhere.

In addition to its use in English-speaking countries, English plays an important role as a technical language around the world, in medicine, computer science, air traffic control, and many other areas of concentrated expertise and formal communication among international professionals.

Grammar

Singular and Plural for Nouns

In British English, singular nouns that describe multiple people are often treated as plural, particularly where one is concerned with the people constituting the team, rather than with the team as an entity. The singular form is usually used in American. For example, British "the team are worried"; American "the team is worried". Americans may use the plural form when the individual membership is clear, for example, "the team take their seats" (not "the team takes its seat(s)"), although it is often rephrased to avoid the singular/plural decision, as in "the team members take their seats". The difference occurs for all collective nouns, both general terms such as team and company and proper nouns (for example, where a place name is used to refer to a sports team). Proper nouns which are plural in form take a plural verb in both American and British English. Examples:

British English: "The Clash are a well-known band." American English: "The Clash is a well-known band." Both: "The Beatles are a well-known band."

British English: "New England are the champions." American English: "New England is the champion." Both: "The Patriots are the champions".

Differences in which nouns are the same in both their plural and singular forms, such as the word sheep. In American English, shrimp is such a word, but in British English the plural of shrimp is shrimps. (Shrimps is occasionally heard in the southern U.S., but is otherwise rare, apart from its colloquial use as a pejorative term for small people). An unusual example is innings, which is both singular and plural in British English, but for which a separate singular form (inning) exists in American English.

Use of Tenses

British English uses the present perfect tense to talk about an event in the recent past and with the words already, just and yet. In American usage, these meanings can be expressed with the present perfect or the simple past.

"Have you cleaned your teeth?" / "Did you clean your teeth?"

"Have you done your homework yet?" / "Did you do your homework yet?"

"I've just got home." / "I just got home."

"I've already eaten." / "I already ate."

Similarly, the pluperfect is occasionally replaced by the preterite in the USA; this is generally regarded as sloppy usage by those Americans who consider themselves careful users of the language.

In British English, have got or have can be used for possession and have got to and have to can be used for the modal of necessity. The forms which include got are usually used in informal contexts and the forms without got in more formal contexts. In American speech the form without got is used more than in Britain. American also informally uses got as a verb for these meanings, e.g. "I got two cars," "I got to go"; but these are nonstandard and will be considered sloppy usage by many American speakers.

The subjunctive mood is more common in American English in expressions such as: "They suggested that he apply for the job". British English would have "They suggested that he should apply for the job" (or even "They suggested that he applied for the job"). However, the British usage ("should apply") is also heard in the United States.

Verb Morphology - The list of Irregular Verbs

Verb past tenses with -ed: American dreamed, leaped, learned, spelled; Commonwealth dreamt, leapt, learnt, spelt. As with the "tre" words, the t endings are occasionally found in older American texts. The forms with -ed are also common in Commonwealth usage, and preferred by many careful writers of English since they are weak verbs

Other verb past tense forms: American fit, forecast, knit, lit, wed; Commonwealth fitted, forecasted, knitted, lighted, wedded. The distinction is, however, not rigorous as the Commonwealth forms are also found in American, and both lit and forecast are not only found in Commonwealth English, but are generally considered standard usage.

Also, the past participle gotten is rarely used in modern British English (although it is used in some dialects), which generally uses got (as do some Americans), except in old expressions such as ill-gotten gains. Commonwealth usage retains the form forgotten, though. Furthermore, according to the Compact Oxford English Dictionary, "The form gotten is not used in British English but is very common in North American English, though even there it is often regarded as non-standard." In North America, most people who use gotten also use got, with gotten emphasizing the action of acquiring, and got tending to indicate simple possession. (Have you gotten it?, as an example)

The past participle proven is frequently used in American English, although some speakers avoid it, and it remains proved in British English (except in adjectival use; and usage is different in Scots law).

American English further allows other irregular verbs, such as thrive (throve–thriven) or sneak (snuck), which remain regular in Commonwealth English, and often mixes the preterite and past participle forms (spring–sprang (U.S. sprung)–sprung), sometimes forcing verbs such as shrink (shrank–shrunk) to have a further form, thus shrunk–shrunken. (The Associated Press Stylebook in American English treats some irregular verbs as colloquialisms, insisting on the regular forms for the past tense of dive, plead and sneak.)

Presence or absence of syntactic elements

Where a statement of intention involves two separate activities, it is acceptable for speakers of American English to use to go plus bare infinitive. Speakers of British English would instead use to go and plus bare infinitive: thus where a speaker of American English might say "I'll go take a bath", British English speakers would say “I'll go and have a bath”. (Both can also use the form “to go to” instead to suggest that the action may fail, as in “He went to take/have a bath, but the bath was full of children.”) Similarly, to come plus bare infinitive is acceptable to speakers of American English, where speakers of British English would instead use to come and plus bare infinitive: thus where a speaker of American English might say Come see what I bought, British English speakers would say, Come and see what I've bought (notice the addition of “have”: a common British preference).

Use of prepositions before days denoted by a single word. Where British people would say “She resigned on Thursday”, Americans often say “She resigned Thursday”, but both forms are common in American usage. Occasionally, the preposition is also absent when referring to months: “I'll be here December” (although this usage is generally limited to colloquial speech).

In Britain, from is used with single dates and times more often than in the United States. Where British speakers and writers may say “the new museum will be open from Tuesday,” Americans always say “the new museum will be open starting Tuesday” or “the new museum will open Tuesday”, and would probably be confused by the British saying. (This difference does not apply to phrases of the pattern from A to B, which are used in both British and American English.)

Where an American will meet with someone, a British person can meet someone. To a British person, the use of with seems to reinforce a notion of mutuality (in this case, the usage is similar to to have a meeting with someone, thus meet with is acceptable, but visit with sounds very strange (you cannot visit someone while they are visiting you).

British people say “I'll write to you” where Americans commonly say “I'll write you”.

Intransitive verbs often become transitive in American

English; for example, British English: "The workers protested against the decision." American English: "The workers protested the decision."

A few "institutional" nouns take no definite article when a certain role is implied: for example, at sea [as a sailor], in prison [as a convict]. Among this group, Commonwealth English has in hospital [as a patient] and at university [as a student], where American English requires in the hospital and at the university. (A nurse, visitor, etc. would be in the hospital in both systems.)

American English distinguishes in back of [behind] from in the back of; the former is unknown in Britain and liable to misinterpretation as the latter. Both however distinguish in front of from in the front of.

American legislators and lawyers always use the preposition "of" between the name of a legislative act and the year it was passed, while their British equivalents do not. Compare Americans with Disabilities Act of 1990 to Disability Discrimination Act 1995.

Different prepositions in certain contexts

In the United States, the word through can mean "up to and including" as in Monday through Friday. In Britain Monday to Friday, or Monday to Friday inclusive is used instead; Monday through to Friday is also sometimes used. (Note: In some parts of Northern England the term while can be used in the same way, as in Monday while Friday.)

In American English on is usually used to say which street one lives in whereas British English usually uses in, eg on Churchill Street /in Churchill Street.

After talk American can use the preposition with but British always uses to (i.e. "I'll talk with Dave / I'll talk to Dave". The American form is sometimes seen as more politically correct in British organisations, inducing the ideal of discussing (with), as opposed to lecturing (to). This is, of course, unless talk is being used as a noun, for example: "I'll have a talk with him" in which case this is acceptable in both British and American English.

In American English from is the correct preposition to use after the word different: "American English is different from British English in several respects." While considered technically incorrect, different than is also commonly heard in the US. The British phrasing different to, however, is almost entirely unknown in the US. Note that, when grammar is taught formally in Britain, both different than and different to are strictly prohibited, whereas different from is considered correct.

Miscellaneous grammatical differences

In American English there is a normative rule which says that, for a non-human antecedent, which should not be used in restrictive relative clauses.

In names of American rivers, the word river usually comes after the name (for example, Colorado River), whereas for British rivers it comes before (as in River Thames). One exception present in British English is the Fleet River, which is rarely called the River Fleet by Londoners outside of official documentation. An exception in the U.S. is the River Raisin in Michigan.

In the names of counties of the British Isles, the word county comes before the name e.g. County Limerick (Ireland) or is simply absent, i.e. Glamorgan (Wales) or Kent (England), whereas in the U.S., it comes after the name, e.g. Washington County.

In British English the word sat is often colloquially used to cover sat, sitting and seated: "I've been sat here waiting for half an hour." "The bride's family will be sat on the right side of the church." This construction is not often heard outside Britain. In the 1960s, its use would mark a speaker as coming from the north of England but by the turn of the 21st century this form had spread to the south. Its use often conveys lighthearted informality, as many speakers intentionally use an ungrammatical construction they would probably not use in formal written English.

This colloquial usage is widely understood by British speakers. Similarly stood can be used instead of standing. To an American these usages may imply that the subject had been involuntarily forced to sit or stand.

In most areas of the United States, the word with is also used as an adverb: "I'll come with" instead of "I'll come along". However, in some British Dialects, 'come with' is used as an abbreviation of 'come with me', as in "I'm going to the office - come with" instead of "I'm going to the office - come with me" (the literal German translation "Komm mit" is occasionally heard).

Word Derivation and Compounds

Directional suffix -ward(s): English forwards, towards, rightwards, etc.; American forward, toward, rightward. The forms with -s are only used as adverbs or prepositions. In American English, the -s forms are fast disappearing, except afterwards. In British English, there is a semantic difference in the usage of the two possible forms. The Oxford English Dictionary states the following about forward and forwards: "/.../ the latter expresses a definite direction viewed in contrast with other directions. In some contexts either form may be used without perceptible difference of meaning; the following are examples in which only one of them can be used: 'The ratchet-wheel can move only forwards'; 'the right side of the paper has the maker's name reading forwards'; 'if you move at all it must be forwards'; 'my companion has gone forward'; 'to bring a matter forward'; 'from this time forward'."

Commonwealth English allows agentive -er and attributive -ing suffixes for football (also cricket; often netball; occasionally basketball). American English always uses football player rather than footballer. Where the sport's name is usable as a verb, the suffixation is standard in both American English and British English: for example, golfer, bowler, and shooter.

English writers everywhere occasionally make new compound words from common phrases; for example, health care is now being replaced by healthcare on both sides of the Atlantic. However, American English has made certain words in this fashion which are still treated as phrases in most Commonwealth countries. For example, Americans write "trademarks," but some other countries write "trade-marks" or "trade marks."

In compound nouns of the form <verb><noun>, sometimes American English favours the bare infinitive where British English favours the gerund. Examples include: jump rope / skipping rope; racecar / racing car; rowboat / rowing boat; sailboat / sailing boat. In each of these pairs, the former term is more common in America than Britain and the latter more common in Britain than America (although it is not necessarily the case that the former is more common than the latter within America or the latter more common than the former within Britain).

Singular attributives in one country may be plural in the other, and vice versa. For example, Britain has a drugs problem while the United States has a drug problem (although the singular usage is also commonly heard in Britain).

General trends

While the use of American expressions in English is often noted in Britain, movement in the opposite direction is less common. But recent examples exist, including the idiom "to go missing," which had been a distinctively British expression but is used increasingly in American English, at least in journalism. The noun "queue" also seems to be making inroads in the U.S. as well. (The usual American equivalents of "to go missing" and "queue" are "to disappear" and "line", respectively.) Also, the spelling of 'travelling' and 'traveller' and 'judgement' through observations appear to be creeping into American English.

Words used only in British English

British English terms, such as lorry, biscuit, chap, and shag although they would not generally use them, or may be confused as to whether one means the American or British meaning of some (such as biscuit). They will be able to guess approximately what is meant by some others, such as candy floss or driving licence. However, use of many other British words, such as naff (unstylish) or busk (to play a musical instrument in public with the hope of getting donations from passers-by), risks rendering a sentence incomprehensible to most Americans.

Words used only in American English

Speakers of British English are likely to be aware of some American English terms, such as sidewalk, gas, cookie, elevator although they would not generally use them. They will be able to guess approximately what is meant by some others, such as cotton candy. However, use of some other American words risks rendering a sentence incomprehensible to most British people.

Note that, largely through the influence of Hollywood, the chance of a given Americanism being understood by a British person is significantly greater than in the reverse case. However, Americans who watch britcoms and other imported BBC programs and British news re-broadcasts on American public television (PBS) are more likely to be familiar with British usages than those who do not.

Fitted is used in both conventions as an adjective ("fitted sheets" are the same size as the mattress) and as the past tense of fit ("to suffer epilepsy", for example, "Leavitt fitted" in The Andromeda Strain); however fit and fitting do not denote epileptic seizure in ordinary British use (though that usage is common within medical circles), as the same effect is achieved by to have a fit or to throw a fit.

Words with differing meanings

In Southern Britain the word whilst is used almost interchangeably with while and whilst is the more common term. Whilst is more often used in instruction manuals, legal documents, etc. To Americans the word whilst, in any context, seems very archaic and/or pretentious. The words amidst (as opposed to amid), and to a lesser extent amongst (as opposed to among) are also rarer in American English. ("In the midst" is a standard idiom in both).

In Britain generally the term "fall" used in the context of the seasons became obsolete and was replaced by the word "autumn". Although archaic, found often in Elizabethan and Dickensian literature, understanding of the word is usually ascribed to its continued use in America.

In Britain the term "period" for a "full stop" is now obsolete, except when used as a phrase, such as "Don't do that. Period." (although this in itself is very likely to be an American import).

Numbers

When saying or writing out numbers, the British will insert an "and" before the tens and units, as in "one hundred and sixty-two" and "two thousand and three", whereas Americans will typically drop the "and" as in "two thousand three"; however, "two thousand and three" is also common. Americans are more likely than the British to read numbers like 1,234 as "twelve thirty-four", instead of "one thousand, two hundred and thirty-four" unless discussing the year 1234, when "twelve thirty-four" would be the norm. Similarly, for the house number (or bus number, etc) "272" British people would tend to say "two seven two" while Americans would tend to say "two seventy-two". Between 1100 and 1900 the British commonly read numbers ending in round hundreds as, for instance, "sixteen hundred" instead of "one thousand six hundred", but from 2000 upwards usage like "thirty-two hundred" would be replaced by "three thousand two hundred".

There was also a historical difference between billions, trillions, and so forth. Americans use "billion" to mean one thousand million (1,000,000,000), whereas in Britain, until the latter part of the 20th century, it was almost exclusively used to mean one million million (1,000,000,000,000), with one thousand million sometimes described as a "milliard", the definition adopted by most other European languages. However, the "American English" version has since been adopted for all published writing, and the word "milliard" is obsolete in English, as are billiard (but not billiards), trilliard and so on.

Nevertheless, the majority of people have no direct experience with manipulating numbers this large, so a significant proportion of international readers will interpret "billion" as 1012, even if they are young enough to have been taught otherwise at school. For this reason, defining the word may be advisable when writing for the general public. However, all major British publications and broadcasters, including the BBC, which long used "thousand million" to avoid ambiguity, now use "billion" to mean thousand million.

Finally, when referring to the numeral 0, British people would use "zero", "nought", or "oh" normally, or "nil" in instances such as sports scores and voting results. Americans use the term "zero" most frequently; "oh" is also often used, and occasionally slang terms such as "zilch" or "zip". Phrases such as "the team won two-zip" or "the team leads the series, two-nothing" are heard when reporting sports scores. The digit 0, for example, when reading a phone or account number aloud, is nearly always pronounced "oh" in both languages for the sake of convenience.

When reading numbers in a sequence, such as a telephone or serial number, British people will use the terms double or triple/ treble. Hence 007 is "double oh seven". Exceptions are the emergency telephone number 999, which is always "nine nine nine" and the apocalyptic "Number of the Beast" which is always "six six six". The directory enquiries prefix 118 is also "one one eight" due to its extensive advertising campaign with the slogan read out as "One one eight, what's your number?". In the US, 911 (the US emergency telephone number) is almost always read "nine-one-one", while 9-11 (September 11, 2001) is usually read "nine-eleven".

There are also variations in floor numbering between the U.S. and U.K. — in most countries, including Britain, the "first floor" is one above the entrance level while the entrance level is the "ground floor"; whereas normal American usage labels the entrance level as the "first floor" and does not use "ground floor." Some American buildings have a "ground floor" or another name for the entrance level, usually as part of a plan to cater to cosmopolitan persons. Nonetheless, the rest of the floors are numbered in the usual American manner.

Figures of speech

Both English and American English use the expression "I couldn't care less" to mean the speaker does not care at all. In American English, the phrase "I could care less" (without the "n't") is synonymous with this in casual usage. Intonation no longer reflects the originally sarcastic nature of this variant, which is not idiomatic in British English and might be interpreted as anything from nonsense to an indication that the speaker does care.

In both areas, saying "I don't mind" often means "I'm not annoyed" (for example, by someone's smoking), while "I don't care" often means "the matter is trivial or boring". However, in answering a question like "Tea or coffee?", if either alternative is equally acceptable, an American may answer "I don't care", while a British person may answer "I don't mind". Either sounds odd to the other.

In American English, "to luck out" means to succeed by

amazingly good luck, whereas in Commonwealth English "to luck out" is a recent colloquialism which means to have no success. Thus, the phrase "I lucked out on getting the free tickets" means either getting or not getting the free tickets, respectively. In some parts of America though, both meanings can and will be understood. The meaning is derived through context.

Business

In his history of the Second World War, Winston Churchill records that differences in the interpretation of the verb "to table" caused an argument between British and American planners. The British wanted a matter tabled immediately because it was important, and the Americans insisted it should not be tabled at all because it was important. In British English, the term means "to discuss now" (the issue is brought to the table), whereas in American English it means "to defer" (the issue is left on the table).

In a similar vein, the verb "to slate" means "to schedule" in the U.S. but (informally) "to disparage" in the UK. Thus a headline such as "Third Harry Potter Film Slated" has two very different interpretations.

One usage of the word "bomb" causes similar confusion: in the U.S. "the show bombed" means it was a total failure; in the UK "the show went down a bomb" means it was a great success. The American slang phrase "the bomb," however (perhaps inspired by African American Vernacular English), almost always indicates positivity. For example, the phrase, "That show was the bomb," would mean that the show was outstanding. In this usage, the article "the" will sometimes be pronounced "da", and sometimes would be written "That show was da bomb", indicating a dialect pronunciation of the word "the".

Education

In the UK, a student is said to "read" or to "study" a subject, while in the U.S., a student either "studies" the subject or "majors" in it. The latter refers only to the student's principal course of study, while the former may be refer to any class being taken.

British English:

"She read history at Oxford".

American English:

"She majored in history at Yale."

In the UK, a student "revises" or "does revision" for an examination, while in American English, the student "studies" for it. When "taking" or "writing" the examination, a student in the UK would have that examination supervised by a "invigilator" whereas in American English it would be a "proctor" (or merely "(exam) supervisor").

In the UK, a student is said to "sit" or "take" an exam, while in the U.S., a student "takes" an exam. In the UK, a teacher "sets" an exam, while in the U.S., a teacher "writes" or "gives" an exam. The expression "he sits for" an exam also arises in British English, but only rarely in American English; American lawyers-to-be "sit for" their bar exams, and American master's and doctoral students may "sit for" their comprehensive exams, but in nearly all other instances, Americans "take" their exams.

British English:

"I sat my Spanish exam yesterday."

"I plan to set a difficult exam for my students, but I haven't got it ready yet."

American English:

"I took my exams at Yale."

"I spent the entire day yesterday writing the exam. At last, it's ready for my students."

Another source of confusion is the different usage of the word college.

In the U.S., this refers to a post-high school institution such as a university, whilst in the UK and most Commonwealth countries it refers primarily to a tertiary institution between high school and university (normally referred to as a "Sixth Form College" after the old name in secondary education for Years 12 and 13, the "6th form") where intermediary courses such as A Levels or NVQs can be taken and GCSE courses can be retaken, with the interchangeability of college with high school being rare but not unknown. Americans may be surprised to hear of a 14 year old attending college in the UK, mistakenly assuming it is at the university level. In both the U.S. and UK, college can refer to some division within a university such as a "college of mathematics and science". Institutions in the U.S. that offer two to four years of post-high school education often have the word college as part of their name, while those offering more advanced degrees (for example, masters, doctorate) are called a university. However, Americans attending either a college or university are often collectively called "college students", and the institutions themselves "colleges", regardless of their status. The words freshman (or the gender neutral term frosh), sophomore, junior and senior refer to the first, second, third, and fourth year respectively of both high school and college (university) students in the U.S. It is important that the context of either high school or college first be established, or else it must be stated directly (that is, "She is a high school freshman." "He is a college junior."). Many institutions in both countries also use the term "first-year" as a gender-neutral replacement for "freshman".

In the UK, first year university students are often called freshers; however, there are no specific names for those in other years, nor for school pupils.

In the UK, the US equivalent of a high school is often referred to as a secondary school regardless of whether it is public or private. In addition, a public school in the UK is, in fact, a private school, while the UK equivalent to public school is a state (funded) school.

Transport

Americans refer to transportation, while British people refer to transport.

Differences in terminology are especially obvious in the context of roads. The British term dual carriageway, in American parlance, would be a divided highway. Central reservation on a motorway in the UK would be a median on a freeway or expressway in the U.S. The one-way lanes that make it possible to enter and leave such roads at an intermediate point without disrupting the flow of traffic are generally known as slip roads in the UK, but U.S. civil engineers call them ramps, and further distinguish between on-ramps (for entering) or off-ramps (for leaving). When American engineers speak of slip roads, or slip ramps, they are referring to on-ramps and off-ramps that have been rearranged (through use of a grade separation) to minimize weaving on a freeway segment between two interchanges that are too close together.

In the UK, the term outside lane refers to the higher-speed passing lane closest to the center of the road, while inside lane refers to the lane closer to the edge of the road; these terms have the opposite meanings in American English, with the outside lane being the one near the edge and the inside lane being the one closer to the median — it is worth noting that Americans also drive on the opposite side to British people, so that the British inside lane is, like the American one, the leftmost one (going in any given direction).

Greetings

In the US, when Christmas is explicitly mentioned in a greeting, the universal phrasing is Merry Christmas. The British alternative, Happy Christmas, is entirely unknown in America, and while its meaning is obvious, it might still garner the speaker some odd looks.

Spelling

Some words shared by all English speakers are spelled one way by Americans (and at times Canadians and Australians) but are spelt differently in some (or, at times, most) other English speaking countries.

Traditionally, many English verbs have been spelled with both -ize and -ise. The -ise is often used over -ize (e.g. organise, legalise) in British English, but -ize is always used in American English (advertise being a rare exception). The American usage is the result of several reforms spearheaded by Noah Webster in the 19th century, while the British usage (where -ize was traditionally dominant) comes from 19th-century French influence. Perhaps because of this, -ize is preferred by Oxford University Press, is listed first in most British Dictionaries, and is preferred by many careful users of English in England; see Oxford's rationale on the matter for more information. The same is true of nouns and adjectives derived from such verbs: organisation vs. organization, etc. One notable exception is the verb to analyse, which although spelled to analyze in American English was never spelled that way in British English. Furthermore, not all such words are spelled with -ize even in American English: for example, incise, circumcise.

Punctuation

Full stops/Periods in abbreviations: Americans tend to write "Mr.", "Mrs.", "St.", "Dr." etc., while British will usually, but not always, write "Mr", "Mrs", "St", "Dr" (or even "D'r"), etc., following the rule that a full stop is used only when the last letter of the abbreviation is not the last letter of the complete word. However, many British writers would tend to write other abbreviations without a full stop, such as "Prof", "etc", "eg", and so on (so recommended by some Oxford dictionaries). The rationale behind this usage is that it is typographically more elegant, and that the omitted full stops/periods are essentially superfluous, as the reader recognizes the abbreviation without them. It also removes ambiguity by reserving the period for ending sentences. However, the "American" usage of periods after most abbreviations is also widely used in Britain. Note that in either case it is incorrect to put a period after units such as kg for kilogram or Hz for hertz, as these are considered unit symbols, not abbreviations; however, the unit for "inch" is properly "in.", as it would be ambiguous without the period.

It is sometimes believed that British English does not hyphenate multiple-word adjectives, such as "a first class ticket". This usage is rare, and often considered incorrect. The most common form is as in American English, such as "a first-class ticket".

Quoting:

Americans start with double quotation marks (") and use single quotation marks (') for quotations within quotations. In general this is also true of British English but can be the opposite when used in book publishing, for example. In journals and newspapers, quotation mark double/single use depends on the individual publication's house style.

Contents of quotations: Americans are taught to put commas and periods inside quotation marks, whereas British people will put the punctuation inside if it belongs to the quote and outside otherwise. This means that direct speech retains punctuation inside inverted commas in British English also, with a full stop changing into a comma if followed by explanatory text.

Carefree means "free from care or anxiety." (American style)

Carefree means "free from care or anxiety". (British style)

"Hello world," I said. (both styles)

The American style was established for typographical reasons, having to do with the aesthetics of commas and quotation marks in typeset text. It also usefully eliminates the need to decide whether a period or comma belongs to the quotation. However, many people find the usage counterintuitive. Hart's Rules and the Oxford Dictionary for Writers and Editors call the British style "new" or "logical" quoting; it is similar to the use of quotation marks in many other languages (including Spanish, French, Italian, Catalan, Dutch, and German). This "logical" style is increasingly popular in America, although formal writing generally calls for the "American" style. In fact, the British style is often the de facto standard among Americans for whom formal or professional writing is not a part of their daily life; many are in fact unaware that the normative American usage is to place commas and periods within the quotation marks. (This rule of placing all punctuation inside quotation if and only if it belongs to the quotation is expressly prescribed by some American professional organisations such as the American Chemical Society; see ACS Style Guide.) According to the Jargon File, American hackers have switched to using "logical" British quotation system, because including extraneous punctuation in a quotation can sometimes change the fundamental meaning of the quotation. More generally, it is difficult for computer manuals, online instructions, and other textual media to accurately quote exactly what a computer user should see or type on their computer.

The American rule was used worldwide up to around the Second World War, after which the more logical style of presenting quotation marks took hold everywhere except America.

Letter-writing:

American students in some areas have been taught to write a colon after the greeting in business letters ("Dear Sir:") while

British people usually write a comma ("Dear Sir,") or make use of the so-called open punctuation ("Dear Sir"). However, this practice is not consistent throughout the United States, and it would be regarded as a highly formal usage by most Americans.

Titles and Headlines

Use of capitalisation varies.

In American English, a capital letter is typically used after an exclamation point, as in: "Good grief, Jim! Leave the man alone." In British English the expression would usually be rendered as 'Good grief, James! leave the man alone.'

Sometimes, the words in titles of publications, newspaper headlines, as well as chapter and section headings are capitalised in the same manner as in normal sentences (sentence case). That is, only the first word is capitalised, along with proper nouns, acronyms, etc.

However, publishers sometimes require additional words in titles and headlines to be capitalised, for added emphasis, as it is often perceived as appearing more professional. In American English, this is common in titles, but less so in newspaper headlines. The exact rules differ between publishers and are often ambiguous; a typical approach is to capitalise all words other than short articles, prepositions, and conjunctions. This should probably be regarded as a common stylistic difference, rather than a linguistic difference, as neither form would be considered incorrect or unusual in either Britain or the U.S. Many British newspapers use fully capitalised headlines for impact, as opposed to readability (examples include The Sun, The Daily Sport, News of the World).

Dates

Date formats are usually written differently in the short (numerical) form. Christmas Day 2000, for example, is 25/12/00 in Britain and 12/25/00 in the U.S., although occasionally other formats are encountered, such as the ISO 8601 2000-12-25, popular among programmers and others seeking to avoid ambiguity. The difference in short-form date order can lead to misunderstanding. For example, 06/04/05 could mean either 4 June 2005 (U.S.) or 6 April 2005 (Britain).

When writing long-form dates, the format "December 25, 2000" is generally encountered in the U.S., and widely encountered in the UK. However, the British are more likely than Americans to use the format "25 December 2000" although it is acceptable in the U.S., and the American grammarians Strunk and White, among others, recommend it. It is common in Britain, and somewhat less so in the U.S., to add a superscripted 'st, nd, rd' or 'th' to the day number in informal writing (thus "25th December 2000" or "December 25th, 2000"), but this is generally regarded as superfluous and avoided in formal use.

Similarly, in American speech, "December twenty-fifth" is the most likely form, though "the twenty-fifth of December" is also not uncommon. In Britain the latter is more likely, and even when the month is presented first the definite article is usually inserted in speech, thus "December the twenty-fifth

— American Idioms —

Smell a Rat

How come the front door is open? Didn't you close it before we went shopping? I'm sure I did. I can't understand it. Frankly, I *smell a rat.*Me, too. I'm *convinced that something is definitely wrong* here. We'd better call the police.

Go to the Dogs

Have you seen their house lately? It's really *gone to the dogs.*It's true that it has *become run-down and in serious need of repair*, but I'm sure that it can be fixed up to look like new.I guess with a little carpentry work and some paint it could look pretty decent.

Fishy

When the security guard saw a light in the store after closing hours, it seemed to him that there was something *fishy* going on. He called the central office and explained to his superior that he thought something *strange and suspicious* was occurring.

Take the Bull by the Horns

Julie had always felt that she was missing out on a lot of fun because of her clumsiness on the dance floor. She had been putting off taking lessons, but she finally *took the bull by the horns* and went to a professional dance studio for help. She was tired of feeling left out and *acted decisively to correct the situation.*

Let the cat out of the Bag

Bob was going to retire from teaching in June, and the foreign language department was planning on presenting him with some luggage at his retirement dinner. He wasn't supposed to know about it, but someone *let the cat out of the bag.* At the dinner Bob acted surprised, even though someone *had told him what he was getting before the official presentation.*

For the Birds

They went to a poetry reading, but they got bored and restless. As far as they were concerned, it was *for the birds*! They left during an intermission because they found the reading *totally uninteresting and meaningless.*

Straight From the Horse's Mouth

How did you find out that Jill was engaged?I got the information *from a very reliable source.*You mean Jill told you so herself?That's right. I got it *straight from the horse's mouth!*

Horse Around

Did you hear about Dave's back injury?No. How did he get hurt?Well, after the coach left the gym he decided to stay and *horse around* on the parallel bars. He somehow lost his grip and fell on his back.That's too bad, but he shouldn't have been *aimlessly playing around* on the equipment without proper supervision.

Cat Got Your Tongue?

Come on, Connie! Tell us what you think about our little ride down the rapids yesterday.Well, uh...Wasn't it exciting? I, uh...What's the matter? *Cat got your tongue*?If you must know, I'm *keeping quiet* because I was scared out of my wits!

Get in Someone's hair

Children! Would you please stop making so much noise!And for heaven's sake, pick up your clothes and toys!It's hard enough trying to keep this house clean without your throwing your things all over the place!Clara, I know that the children *get in your hair*, but you should try not to let it upset you so much.Listen, Jim. I can't help it . The children *bother me and make me very angry* when they're so noisy and messy.

Shoot Off One's Mouth

Jim doesn't play tennis very much, but he's always *shooting off his mouth* about how good he is. Yet he's fooling nobody. Jim is somewhat of a braggart and everyone knows that he gives opinions without knowing all the facts and *talks as if he knew everything about the game.*

Jump Down Someone's Throat

That's it, Greg! You'd better not come in after midnight again tonight!I know, dad. You don't have to *jump down my throat*! I told you that I'd make it home around 11:30. I don't intend to be late!Well, you've said that before and in you come at 2: 30 in the morning. You can't blame me for *getting angry and scolding you.* I've got good reason.

Pay Through the Nose

At last Mr. Smith came upon the rare stamp he had been seeking at an auction. Since many other stamp collectors would also be bidding for it, he realized that he would have to *pay through the nose* in order to have it.After considering the increasing value of the stamp, he decided that he would not mind *paying such a high price* for something so rare.

Tongue-in-Cheek

Why were you teasing Sonia about her new hairdo? She really took offense at what you said.I didn't mean to offend her. I was simply making a *tongue-in-cheek* remark when i said that it was too elaborate for a girl of her young, tender age.Well, she thought you were serious. She had no idea that you were just saying that *as a joke*.I'm really sorry. I suppose I owe her an apology.

Pull Someone's Leg

Hey, Al. I was invited to be a judge for the Miss America Beauty Pageant!Oh, really? Come on, you're *pulling my leg*!No, honestly. Do you really think that I'm trying to *fool you with a ridiculous story*?Well, you've told me foolish stories before.I can assure you that this one is for real.

Play It by Ear

Let's go to the movies, agreed? Sure. And what'll we do after that ?Oh, I don't know. Let's *play it by ear*.Well, I would like to have a more definite plan of action.Don't be like that. It's always more fun *not knowing what to expect and deciding what to do as we go along.*

Stick Out One's Neck

How come they're asking me to act as their guide through the jungle?Evidently they think you're the only one who can lead them to the lost temple.That jungle has danger lurking around every corner.Why should I *stick my neck out* for them? They didn't pay me for my services.They know that you would be *taking a great risk and could possibly get hurt*, but you're the only one with enough knowledge to take them to their destination. I'm sure you'll be amply rewarded.

All Thumbs

I guess I'm *all thumbs* because I'm so nervous. I'm already late for my date.Well, I suppose that being so nervous would make you *clumsy and awkward*. But don't worry. I'm sure your date will wait.

Not Have a Leg to Stand On

Tom maintains that the firm owes him some back wages for having worked overtime. However, he *won't have a leg to stand on* unless he can prove that he put in all those extra hours. He doesn't stand a chance of getting his money *without a strong foundation of facts to support his position.*

Get Off Someone's Back

Hey, John. I'm bored. Come on, let's go out and do something.Sorry, I'm right in the middle of studying for a physics exam. I won't be able to make it tonight.You've been studying for a long time. Why don't you take a break ? Come on! Let's go ! Forget studying for a while !Look! *Get off my back*! I can't go anywhere!OK. I'll *stop bothering you* only if you promise to let me know the minute you're finished.

Drive Someone Up a Wall

Wow ! What a great set of drums!Yeah, they're great, but I can't play on them when my folks are at home. They say *I drive them up a wall* with all the loud banging.I get the same thing at home. My folks tell me that I *annoy them and get them really angry* whenever I turn up the volume on my stereo.

String Someone Along

Liz had high hopes of marrying Dean. When he ran off with another woman, she realized that he was just *stringing her along*. She had felt very strongly about him and was really hurt to see that he was *deceiving her* and had no intentions of ever marrying her.

Leave Someone High and Dry

Say, Jill. I thought that John was going to help you do the

dishes tonight.So did I. But he *left me high and dry.*Where did he go ?Well, he got a call from some of his pals at work to go bowling, and he *left me alone to do all this work without any help at all!*

Spill the Beans

Did you know that Harry was going to take Kathy on a Caribbean cruise ?Yes, I did. He was planning on surprising her with the tickets fro their anniversary, but someone *spilled the beans.*What a shame ! That was supposed to have been a surprise.Yes, it's too bad that someone *told her about the trip beforehand* and ruined Harry's surprise.That's OK. Her enthusiasm was not dampened in the least!

Bite the Dust

Andy did exceptionally well in all of the track events, but he *bit the dust* in the high jump competition. Much to the disappointment of his fans, he *went down in defeat, losing to a competitor* from the visiting team.

Bend Over Backwards

When Joan first started teaching she was afraid that she would have a lot of trouble getting used to the kids and to the faculty. Her fears turned out to be unfounded, since everybody *bent over backwards* to help her. Everyone *tried very hard* to help her feel comfortable and adjust to the school.

Cough Up

Say, Greg. Did you finally get that computer that you wanted so much ?Not yet. I needed to raise a couple of hundred dollars more.Is it going to take you a while to raise the money?It would have taken me forever, but dad said he'd *cough up* the money I need since I'm going to be using the computer for my school work.Maybe it was *difficult for your dad to give* you the money but then, he knows that it's for a good cause.

Jump the Gun

Denise was planning on telling her grandparents that the

doctor said she was going to have twins, but when her dad found out he *jumped the gun* and told them before Denise could say a word. He was so excited that he *became hasty* and revealed the news before Denise had a chance to tell them.

Scratch Someone's Back

Hey, Bea. I need some help stacking these boxes. Would you please give me a hand?OK. And I need some help tidying up the house. How about your helping me out after that ? OK. If you *scratch my back, I'll scratch yours.*I know you don't like doing housework, but I'll *help you with the boxes if you promise to return the favor.*No problem. I'll even do the windows.

Turn Someone Off

How was your date with Marty last night ?Well, it started of OK, but he really *turned me off* when we went for a snack after the movies.Did he say or do something to annoy you?Frankly, he *disgusted me* when he tried to talk with his mouth full.I don't blame you. That would have really bothered me too.

Kick the Bucket

It's been said that the old man knew of a buried treasure, but he *kicked the bucket* before telling anyone where it was. If the treasure exists, the old man unfortunately took the secret of its location with him when he *died.*

Wet Blanket

James was not invited to go on the outing with the rest of the group because he's such a *wet blanket.* On many previous occasions he has *kept others from enjoying themselves by his pessimism and lack of enthusiasm.* It's understandable that no one wants him around.

Up One's Sleeve

All right, Sara. We know that you're planning something big for Jean's birthday. Mind telling us just what you have *up your sleeve* ? I wanted to make his birthday a very special event.Jean has a sister living in France, and I sent her an airplane ticket so that she could be here for his birthday.Boy! That is something special. We kind of guessed that you had some *concealed plan* and were waiting for the right time to reveal it.Well, I didn't want to say anything until I was sure she could come.

Knock Someone's Socks Off

Hi, John. What's new ?Oh, nothing too much with me, but you ought to see Fredo's new car. It'll *knock your socks off*! So, he finally got that Italian sports car he's been dreaming about.He sure did. When you see all the custom features that it has, you'll *get so enthused and excited* you won't know what to do.Boy, I can hardly wait to go for a ride in it.

Lose One's Shirt

I happened to bump into Dole at lunch yesterday afternoon.What's new with Dole these days ?He wasn't doing so well. For one thing, he told me he *lost his shirt* at the races.Dole has always liked to bet on the horses. I'm not surprised that he *lost a great deal of money*.Yeah. At this rate he'll never have a penny to his name.

Lemon

Have you seen Joe's new car yet ?Yeah. It looks good, but she's had nothing but problems with it.That's too bad. It sounds like she got a real *lemon*.She sure did. No sooner did she drive it home from the dealer's than it proved *defective* and started breaking down.

Out of the Woods

Although Eric was well on his way to recovering from his bout with pneumonia, he was still not *out of the woods*.The doctors told him that he would have to take it easy and avoid exposure to cold, since he was not *out of danger and difficulty yet*.

Eating Someone

Hey, Alice. What's been *eating you* lately? Don't you realize how rude and irritable you've become?I know. I'm really sorry for the way I've been acting.Well, why don't you tell me what has been *bothering and upsetting you* and maybe we can work your problem out together.I'll admit that it would help to talk to someone about it.

On the Line

Lately Tom's been more conscientious about the accuracy and quality of his work with the company. He was warned that his job was *on the line* because of his lack of concern for his duties. When Tom was alerted that he was *in danger of losing his job*, he began to take his obligations with the company more seriously.

Feel Like a Million Dollars

I bumped into Nick at the salon yesterday. He looked great, but I noticed that he had a slight limp when he walked.I guess you didn't know that he had an operation on his knee.No, I didn't. How's he feeling ?He says he's *feeling like a million dollars* now. Apparently the pain in his knee is all gone.It's good that he's *feeling so wonderful.* It must be a refreshing change not having to put up with all that discomfort.

Hang On

During the depression years the Smiths had a great deal of trouble with their business, but somehow or other they were able to *hang on.* Although they almost lost their store, they managed to *persevere* until things got better.

Make Ends Meet

It's almost impossible trying to keep up with the high cost of living.It's true. Things are so expensive nowadays that it's very difficult to make ends meet.You know, even with Lucie's salary, our combined income is *hardly enough to pay all the bills.*

Money talks

We've been waiting for three months to get delivery on our

car, and people who put in their order after us have already gotten theirs.Well, *money talks*. Why don't you try giving the dealer a little something extra to move things along ?I know full well that *money has the power to influence people*, but I refuse to pay extra for a service that is owed to me as a client.If you want to have your car maybe you'd better reconsider.

The Early Bird Catches the Worm

Marc, the lines for the rock festival are going to be miles long! If you expect to get tickets for you and Mary, remember that old saying. *"The early bird catches the worm"* I guess you're right. Mary is looking forward to the concert, and I'd hate to disappoint her. I'll *get up real early to get a place at the head of the line*.That way I'll get the tickets I want, for sure ! with a command over English language. This is the ideal unskilled labor that the call center industry is looking for.

4
LIST OF CALL CENTERS IN INDIA

24/7 Customer- Customer Interaction Management company in Bangalore, Chennai, Hyderabad.

24x7 BPO-Accounting, Technical Support / IT Helpdesk, Data Entry, Insurance, Chat and Email Support, CAD/GIS in Ahmedabad.

A R Digitech -CAD/GIS company in KANPUR .15 Clients are from US and UK.

A to Zee BPO Services- Main work is Form Processing, Chat Support, Inbound / Outbound voice, Accounts Receivable / Payable / Reconciliation, CAD/GIS.

A Yantram- BPO service inbound / Outbound voice, B2B Lead Generation, Contact Center / Customer Care, Credit Card Processing, Inbound Call Center, Lead Generation in Ahmedabad.

A-ONE Biz Solutions -Inbound / Outbound voice, Knowledge Management, Health Care, Medical Coding Services, Healthcare Claims Processing company in Chennai.

Aaragon BPO Solutions Pvt. Ltd. -Form Processing, Business Transcription Services, Contact Center / Customer Care, Data Conversion, Data Digitization, e-publishing company in Mumbai .

ABP Solutions -e-publishing, Litigation, Medical Billing company in Chennai .

ABPOC System- Accounting, Billing, Data Conversion, Data Entry, Invoice Processing company in Chennai

Abric e Calls(P) Ltd. Contact Center / Customer Care, Collection Services, Loan/Mortgage Processing company in Dehradun.

Abstract eServices (P) Ltd. Data Processing/ Management, Technical Support / IT Helpdesk, Tele-Marketing, B2B Lead Generation, Collection Services company in New Delhi.

Accel LTD -Form Processing, Data Capture & Processing, Medical Transcriptions, Data Conversion, Data Entry, CAD/GIS company in Chennai.

Access Dataline Private Limited.HR, Payroll, Statutory Compliance, Benefit Administration Services (PF/ESIC), Check Processing company in Mumbai

Acclaris Business Solutions Pvt. Ltd.Claims Processing, Credit Card Processing, Transaction Processing, Accounts Receivable / Payable / Reconciliation, Loan/Mortgage Processing, Tax Management company in Kolkata, Vizag

Acme Technologies Data Processing / Management, Claims Processing, Data Digitization, e-publishing, Check Processing, Healthcare Claims Processing

ACMECeeYess Softech Pvt. Ltd. Data Processing / Management, Medical Transcriptions, Data Digitization, Payroll, Data Mining, Data Validation Services, Document Imaging, Check Processing, Medical Billing company in Bangalore Privately owned .

Acuserv Business Processes Private Limited. Technical Support / IT Helpdesk, B2B Lead Generation, Contact Center / Customer Care, Customer Acquisition, Loan/Mortgage Processing, Receivables Management company in Hyderabad

Adaarsh Technologies- Form Processing, Data Capture & Processing, Data Conversion, Data Entry, XML Tagging company in Bangalore.

Adarsh Solutions Pvt. Ltd. Content Management, Data Conversion, Data Digitization, XML Tagging, Data Mining, Recruitment Services company in Bangalore.

ADG Infotek Inc -Technical Support / IT Helpdesk, Tele-Marketing, Inbound Sales, Inbound / Outbound voice, Inbound Call Center, Technical Writing.

Adiance -BPO Technologies Content Management, CRM, Contact Center / Customer Care, Business Services Provision, Content Outsourcing Services company in Ahmedabad.

Adith Multilingual Services Pvt. Ltd. Foreign Language Translations, eLearning Bangalore.

Adithya Infotech Inbound/Outbound voice, Surveys, Structured Finance (Mortgage/Asset backed Securities), Medical Coding Services, Medical Billing company in Secunderabad.

Advanced Investment Mechanics (I) Pvt. Ltd.- Business Feasibility, Business Research For Banking Industry, Business Services Provision, Competitor Analysis, Content Outsourcing Services company in Mumbai.

Adventity Contact Center / Customer Care, Knowledge Management, Transaction Processing, Finance company in Mumbai.

AEL Data Form Processing, Data Capture & Processing, Data Entry, OCR Services, e-publishing, Document Imaging company in Chennai.

Ajuba Solutions (I) Pvt Ltd. Claims Processing, Collection Services, Medical Coding Services, Revenue Cycle Management Services, Medical Billing company in Chennai.

Alden Business Solutions Accounting, Foreign Language

Translations, Data Conversion, Health Care, Litigation, Financial Transcription company in Chennai.

Allsec Tech Email Support, Chat Support, Inbound / Outbound voice company in Chennai.

AllServe Systems India Ltd. Inbound / Outbound voice, B2B Lead Generation, Accounts Receivable / Payable / Reconciliation, Collection Services, Loan/Mortgage Processing company in Gurgaon.

AllZone Management Solutions Accounting & Finance, Data Conversion, Litigation, Healthcare Claims Processing, Medical Billing company in Chennai.

Aloft Systems Pvt. Ltd. Data Processing / Management, Data Conversion, Data Entry, Check Processing company in Trivandrum.

Alpha Thought Claims Processing, Health Care company in Noida.

Alsol Technologies Form Processing, Data Capture & Processing, Data Entry, Data Digitization, Data Services company in Delhi.

Altius Infosystems Pvt Ltd Data Capture & Processing, Data Conversion, Data Entry, XML Tagging, e-publishing, Litigation Chennai.

Amba Outsourcing Data Processing / Management, Data Capture & Processing, Data Conversion, Data Entry, Chat and Email Support, Data Services company in Coimbatore.

Ambal Technologies Form Processing, Data Processing / Management, Data Conversion, e-publishing, Healthcare Claims Processing, Content Outsourcing Services company in Chennai

Ameriteck Global IT Solutionz Contact Center / Customer Care, Collection Services, Chat and Email Support company in Pune.

Amex Information Tech Ltd. Form Processing, Data Processing / Management, Technical Support / IT Helpdesk, Document Management, XML Tagging, Check Processing company in Mumbai.

Amtier InfoTech Ltd. Inbound / Outbound voice, B2B Lead Generation, Contact Center/Customer Care, Credit Card Processing, Billing Collections company in Gurgaon.

Analyze-It Technologies Data Processing / Management, Market Research, Customer Management, Data Mining, Management reporting, Customer Profiling company in New Delhi

Anlage Infotech (India) Pvt. Ltd. HR, Accounting & Finance, Payroll company in, Bangalore.

Anode Technologies Pvt. Ltd. Data Entry company in Mumbai.

Apex IT World Data Processing / Management, Business Transcription Services, Inbound / Outbound voice, B2B Lead Generation, Data Conversion, Real-time Check Processing company in Coimbatore.

Apollo Health Street Pvt. Ltd. Medical Transcriptions, Health Care, Medical Coding Services, Healthcare Claims Processing, Healthcare Payer, Medical Billing company in Hyderabad.

Aress Software & Education Technologies Pvt. Ltd. Technical Support / IT Helpdesk, Contact Center / Customer Care, XML Tagging, Chat and Email Support, Advanced Tech Support company in Nashik.

Arvato Services India Contact Center / Customer Care, Order Processing, Surveys, Remittance & Payment Processing, Receivables Management company in Gurgaon.

Ascent Consulting Services Pvt.Ltd. Accounting & Finance, Accounts Receivable / Payable / Reconciliation, Payroll, Benefit Administration Services (PF/ESIC) company in

Bangalore.

Assent Europe Limited Tele-Marketing, Inbound / Outbound voice, B2B Lead Generation, Lead Generation, Process Outsourcing, Legal Outsourcing company in Gurgaon.

Astron Data Capture & Processing, Business Transcription Services, Insurance Claims Processing, Accounting & Finance, Transaction Processing company in Trivandrum, Chennai.

Atlas Legal Legal Outsourcing, Legal Research Dallas [US], company in Bangalore

Aumenta Services Technical Support / IT Helpdesk, Contact Center / Customer Care, Chat and Email Support company in Mumbai.

Avadis Network Solutions Pvt. Ltd. Data Capture & Processing, HR, Accounting & Finance, Transaction Processing, Payroll, Lead Generation company in Delhi

Avenue Media (P) Ltd. Data Processing / Management, Data Conversion, XML Tagging, e-publishing company in Mumbai.

Avid It Solutions Medical Transcriptions, Inbound / Outbound voice, Network and Security Management, Inbound Call Center, Airline company in Hyderabad.

Axiom Technologies Technical Support / IT Helpdesk, Contact Center / Customer Care, Transaction Processing company in Hyderabad.

Axion Consultancy And Outsourcing (P) Ltd Accounting & Finance, Contact Center / Customer Care, Data Entry, Accounts Receivable / Payable / Reconciliation, Payroll, Billing Collections company in New Delhi

Azilon Corporation Knowledge Management company in Bangalore.

B2K Corp Technical Support / IT Helpdesk, Contact Center

/ Customer Care, Finance, Health Care, Medical Billing company in Bangalore.

Baxy Infosol Technical Support / IT Helpdesk, Tele-Marketing, Accounting & Finance, Insurance, Telecom Sector, Medical Billing company in Gurgaon

Bay Area Credit Service (I) Pvt. Ltd. Collection Services Pune.

Bayinfocomm Pvt. Ltd. Email Support, Inbound / Outbound voice, B2B Lead Generation, Helpdesk Services, Chat and Email Support, Content Outsourcing Services Chandigarh.

Best Praktizes India Accounting, HR, Accounts Receivable / Payable / Reconciliation, Payroll Bangalore.

Better, Inc. B2B Lead Generation, Contact Center / Customer Care, Customer Acquisition, Sales Intelligence, Upselling and Cross-selling Hyderabad, Pune.

Bhawani Infotech Pvt. Ltd. Data Capture & Processing, Data Conversion, Data Entry, OCR Services, XML Tagging Delhi.

Big Pond Technologies India Private Limited. Email Support, Chat Support, Inbound / Outbound voice Trivandrum USA Funded 50 Rogers Wireless, Nextel Corporation, Global Processing.

Billcorp Technical Support / IT Helpdesk, Tele-Marketing, Medical Transcriptions, Remittance & Payment Processing, Receivables Management Jalandhar (Punjab)

Bizprout Corporate Solutions Pvt. Ltd. Data Processing / Management, Accounting & Finance, Accounts Receivable / Payable / Reconciliation, Payroll, Invoice Processing, Corporate Secretarial Services Bangalore.

Bizzsource Contact Center / Customer Care, Collection Services, Lead Generation New Delhi.

BK ONE Corporate Training Pvt. Ltd. Tele-Marketing, CRM, Contact Center / Customer Care, Staffing Solutions and Training, Chat and Email Support Gurgaon, Bangalore, Noida, New Delhi, Jaipur, Chennai, Hyderabad.

Blue Matrix Software Consultancy Form Processing, Medical Transcriptions, Chat and Email Support, Data Services, Medical Billing Nasik.

BPO Info India Data Processing / Management, Data Conversion, Data Entry, e-publishing, Data Mining, Data Services Mumbai.

Brandgain India Pvt. Ltd. Technical Support / IT Helpdesk, Tele-Marketing, Inbound / Outbound voice, Contact Center / Customer Care, Chat and Email Support New Delhi.

P Info Serve Pvt. Ltd. Claims Processing, Medical Coding Services, Health Insurance, Helpdesk Services, Healthcare Payer, Medical Billing Belgaum (Karnataka).

Business Accounting Services Inc Accounting & Finance, Credit Card Processing, Accounts Receivable / Payable / Reconciliation, Check Processing, Offshore Advisory Services Bangalore, London, Manassas, Singapore Angel Investor 63 Confidential.

Business Oriented Systems Ltd. Data Processing / Management, Inbound / Outbound voice, Collection Services Mumbai.

CA InfoTech Medical Transcriptions, Market Research, Data Entry, Data Digitization, OCR Services Madurai.

Caliber Point Business Solutions Ltd. Form Processing, Claims Processing, Document Management, Knowledge Management, Accounts Receivable / Payable / Reconciliation Mumbai.

Calibrated Healthcare Systems OCR Services, Medical Coding Services, Health Insurance, Healthcare Claims

Processing, Healthcare Payer, Medical Billing New Delhi.

Call Logic Outsourcing Process Solutions, Staffing Solutions and Training, Process Outsourcing, Process Consulting Hyderabad, New Delhi.

CallZone, Inc Inbound / Outbound voice, B2B Lead Generation, Presales, Inside Sales, Sales Intelligence, Opportunity Plan HYDERABAD.

Camarilla Infotech Private Limited. eLearning, Inbound / Outbound voice, Contact Center / Customer Care, e-publishing, Health Care, Medical Billing Chennai.

Cambridge Insurance Claims Processing, Accounting & Finance, Inbound / Outbound voice, Transaction Processing, Loan/Mortgage Processing, Process Consulting Bangalore.

Cambridge Systems (India) Pvt. Ltd. Medical Transcriptions, Recruitment Services, Process.

Candor Business Solutions Pvt. Ltd. HR, Accounting & Finance, Payroll, Statutory Compliance, Benefit Administration Services (PF/ESIC), Gratuity and Super Annuation Bangalore.

CAP Digisoft Solutions Pvt. Ltd. Accounting & Finance, Medical Transcriptions, Litigation, Legal Outsourcing Coimbatore

Care Principium Pvt Ltd. Technical Support / IT Helpdesk, Inbound / Outbound voice, B2B Lead Generation, Contact Center / Customer Care, Accounts Receivable / Payable / Reconciliation, Collection Services Pune.

CBay Systems Medical Transcriptions, Document Management, Health Care Mumbai, Bangalore.

Ceequence Technologies Tele-Marketing, Inbound / Outbound voice, B2B Lead Generation, Contact Center / Customer Care, Transaction Processing Chennai 1 million 175 Fortune 500 Insurance company in the UK Telecom service provider in UK Mortgage intermediary.

Cellbion Interactive Pvt. Ltd. Technical Support / IT Helpdesk, Email Support, Chat Support, Inbound / Outbound voice, Transaction Processing Mumbai - 60 US based companies.

Chakkilam Infotech Limited. CRM, Medical Transcriptions, e-publishing, Health Insurance, Navigation charting, CAD/GIS Hyderabad Listed company 320 Indian and US companies Charon Tec Pvt Ltd Data Conversion, Data Digitization, OCR Services, XML Tagging, e-publishing, Document Imaging Chennai.

Cheers Interactive India Pvt. Ltd. Data Capture & Processing, Market Research, Knowledge Management, Data Mining, Data Validation Services, Data Services Navi Mumbai Privately Funded 90 UK based Knowledge Management Company,UK based Business Information Company, UK based Publishing House, UK based Marketing Intelligence Company, US based Home Shopping Company,US based Utility Company **ChennaiBeats** Form Processing, Claims Processing, Data Entry, Data Digitization, Data Services, Medical Billing Chennai.

Cherukuri BPO - Outsourcing BPO India Data Processing / Management, Business Transcription Services, Accounting & Finance, Data Entry, Banking, Application Performance Monitoring Hyderabad.

CitiusTech Content Management, Technical Support / IT Helpdesk, Billing Mumbai.

Cleave Global e-Services Ltd. Market Research, Contact Center / Customer Care, Data Entry, Collection Services, Helpdesk Services Noida.

Client Outsource Technical Support / IT Helpdesk, Email Support, Market Research, Order Processing, Accounts Receivable / Payable / Reconciliation Chennai.

CMC Limited. Document Management, eCRM Solutions,

Knowledge Management, Network and Security Management, Helpdesk Services Mumbai.

Codec Communications Pvt. Ltd. Accounting, HR, Inbound / Outbound voice, Data Entry, Medical Billing Pune Hands On Ventures 160 BAY Area Credit Services.

Colwell & Salmon Communications (India) Ltd. Technical Support / IT Helpdesk, B2B Lead Generation, Market Research, Telecom Sector, Loan/Mortgage Processing New York(2), New Delhi (NCR).

Combine Resources Inbound Sales, Email Support, CRM, HR, Accounting & Finance, Contact Center/Customer Care in New Delhi.

Commerce Clay Tele-Marketing, Contact Center / Customer Care in New Delhi.

Compcare Data Processing / Management, Data Capture & Processing, Data Conversion, Data Entry, Data Digitization, Document Imaging Delhi.

Computech Enterprise Solutions Pvt. Ltd. (CES) Tele-Marketing, HR, Accounts Receivable / Payable / Reconciliation, Payroll, Health Care, Helpdesk Services, Retail Customer Support Hyderabad Own 240 Computech Corporation, IBS, Careorigin.

Computer Solutions Form Processing, Data Capture & Processing, Data Conversion, Data Entry Delhi.

Conceptual Edge Process Solutions, eCRM Solutions, Operations Services, Upselling and Cross-selling, Process Outsourcing Delhi.

Congruent Solutions eLearning, Finance, Retirement Planning Chennai 100 Consultants'R'Us Business Feasibility, Accounting & Finance, Credit Card Processing, Accounts Receivable / Payable / Reconciliation, Check Processing Cochin.

Consulting Geo Info Systems Data Processing / Management, Data Capture & Processing, Data Conversion, Data Digitization, Data Mining, CAD/GIS New Delhi.

Continuum Systems Chat Support, eCRM Solutions Gurgaon - 25 US and European markets

Convergys Billing, Contact Center / Customer Care Bangalore, Gurgaon Listed Company: NYSE 2000

Core Software Pvt. Ltd. Data Processing / Management, Data Capture & Processing, Data Conversion, Data Entry, Data Digitization Hyderabad.

Cosmic Communication Data Digitization, e-publishing, Document Imaging Ahmedabad.

Cosmic Internet Technologies Solutions Pvt. Ltd. Data Processing / Management, Chat Support, Accounting & Finance, Credit Card Processing, Accounts Receivable / Payable / Reconciliation, Tax Management Bangalore.

Cossmoss InfoTech Inbound / Outbound voice Kolkata 5 crore 150 Howarah Mills Co.Ltd. vision India.

Creative Mediapulse Technologies Pvt. Ltd. eLearning Bangalore.

Crestcom Consultant Pvt. Ltd., People Solutions Chennai, Bangalore.

Cross-Tab Marketing Services Pvt. Ltd. Data Processing / Management, Market Research, Surveys, Data Entry, Data Mining Mumbai, Bangalore.

Crossdomain Accounting & Finance, Employer Services Bangalore, Chennai, Delhi, Mumbai.

CRP Technologies [India] Pvt.Ltd. HR, Market Research, Operations Services, Credit Appraisals, Business Research For Banking Industry, Third Party QC/QA Mumbai, Delhi, Pune, Chennai, Bangalore, Hyderabad, Kolkatta, Bhubhaneshwar,

Rajkot, Surat, Baroda, Palghar, Raigad, Nagpur, Aurangabad, Kolhapur, Nasik, Mangalore, Hubli, Mysore, Vizag, Vijayawada, Trichy, Salem, Madurai, Coimbatore, Cochin, Calicut.

Customer First CRM, Inbound Call Center Bangalore, Chennai, Coimbatore, Hyderabad, Kochi, Kolkata, Pune.

CustomerLine Private Limited. Tele-Marketing, CRM, Inbound / Outbound voice, B2B Lead Generation, Billing, Inbound Call Center Kochi.

Cyber Futuristics India Pvt. Ltd. Technical Support / IT Helpdesk, Market Research, Contact Center / Customer Care, Order Processing, Surveys, Credit Card Processing Jaipur.

Cyber Infrastructure Inbound Sales, B2B Lead Generation, Inbound Call Center, Lead Generation, Helpdesk Services, Chat and Email Support Indore.

Client List Cybiz Call International Ltd. Technical Support / IT Helpdesk, HR, Payroll, Lead Generation, Data Validation Services Gurgaon - 150 UK and USA based companies.

Daemon Information Systems Content Management, Data Processing / Management, Data Conversion, e-publishing Ahmedabad' India.

Daksh eServices Pvt. Ltd. Tele-Marketing, Email Support, Chat Support, Inbound / Outbound voice, Knowledge Management, Transaction Processing, Collection Services, Travel, Transportation & Hospitality Industry Bangalore, Chandigarh, Kolkata, Mumbai, New Delhi (NCR), Pune, Manila.

Data Infosys Limited. Email Support, Chat Support, Inbound/ Outbound voice, Contact Center / Customer Care Jaipur.

Data Management Systems Pvt. Ltd. Data Processing / Management, Data Capture & Processing, Document Management, Data Conversion, Data Digitization, XML Tagging Lucknow.

Data Point Computer Services Data Capture & Processing, Accounting & Finance, Document Management, XML Tagging, Banking, Process Outsourcing Vadodara.

Data Solution Labs Data Processing / Management, Data Capture & Processing, Data Conversion, Data Entry, Data Digitization, Data Validation Services Bangalore.

Datamatics Technologies Limited. Content Management, Accounting & Finance, Document Management, e-publishing, Transaction Processing, Accounts Receivable / Payable / Reconciliation, Health Care, Tax Forms Mumbai, Chennai.

Data Smith Solutions Form Processing, Content Management, Claims Processing, Data Conversion, e-publishing, Document Imaging Chennai.

DCS BPO Private Limited. Form Processing, Claims Processing, Data Capture & Processing, e-publishing Chennai.

Deccan Multisourcing Private Limited Accounting, Technical Support / IT Helpdesk, Tele-Marketing, Accounting & Finance, Telecom Sector Hyderabad Deccan Limited, United Kingdom 20 swansys carephonewarehouse twiceware twicecare.

Deepak Software Consultancy Pvt. Ltd. Data Entry MUMBAI N.A 30 N.A.

DeftPro Solutions Private Limited Accounting, Transaction Processing, Process Outsourcing, Tax Management Chennai.

Delta Services India Pvt. Ltd. Data Digitization, Insurance, Banking, Finance, Payroll, Staffing Solutions and Training Mumbai.

Destination IT Solutions Medical Transcriptions, Data Digitization, XML Tagging, Data Services Visakhapatnam.

Destiny Infotek Limited. Data Processing / Management,

Business Transcription Services, Medical Transcriptions, Data Conversion, Data Entry Kolkata.

Dexterity KPO Services Data Processing / Management, Market Research, Market Analytics Chennai.

Dhooran Conversion Solutions & Services Data Capture & Processing, Medical Transcriptions, Data Conversion, Data Entry Erode.

Digital Access Pvt. Ltd. Tele-Marketing, Inbound / Outbound voice, Travel, Lead Generation, Loan/Mortgage Processing Kolkata Can not be disclosed 120 ICICI, Citibank, Global processing

Digital Avenues Technical Support / IT Helpdesk, Accounting & Finance, Inbound / Outbound voice, Contact Center / Customer Care, Collection Services, Chat and Email Support Kolkata.

Digital Compusoft Inc Medical Transcriptions, Document Management, Medical Billing Piscataway, New Jersrey, Vadodara.

Digital Info soft India (P) Ltd. Form Processing, Data Conversion, Data Entry, Data Digitization, Document Imaging Jaipur.

DPS Technologies India Pvt. Ltd. Data Capture & Processing, Insurance Claims Processing, Credit Card Processing, Data Conversion, Data Digitization, Transaction Processing Kolkata, New Delhi, Siliguri, USA, London.

DPSOFT Information Technologies Pvt. Ltd. Form Processing, Data Capture & Processing, Document Management, e-publishing, Third Party QC/QA, CAD/GIS Pune.

Drishti Foreign Language Translations, Content Management, Business Transcription Services, e-publishing, Resume Management Services, Technical Writing Pune, Bangalore.

Dukestar Technologies Pvt. Ltd. Form Processing, Data Processing / Management, Accounting & Finance, Credit Card Processing, Data Mining, Engineering Services Pune.

Dynamic Software Solutions Market Research, Data Conversion, Data Entry, XML Tagging, Litigation, Legal Research Chennai.

e-Call Serve Outsourcing Pvt. Ltd. Tele-Marketing, B2B Lead Generation, Billing, Contact Center / Customer Care, Inbound Call Center, Chat and Email Support Kolkata.

e24 Technologies Technical Support / IT Helpdesk, Business Transcription Services, Inbound / Outbound voice, Contact Center / Customer Care, Chat and Email Support, Medical Billing Chennai.

E2E Serwiz Solutions Ltd. (Tata Sons Subsidiary) Contact Center / Customer Care, Telecom Sector, Customer Interaction Management Pune, Hyderabad , Chandigarh.

E4e Chat Support, Network and Security Management Bngalore.

Earnest John Technologies Limited. Accounting, Accounting & Finance, Accounts Receivable / Payable / Reconciliation, Payroll, Tax Forms, Tax Management Mumbai.

Ebix-India Tele-Marketing, Insurance Claims Processing, Email Support, Chat Support, Accounting & Finance, Inbound / Outbound voice, B2B Lead Generation, Knowledge Management, Insurance, Banking Noida.

Edgesys Accounting & Finance, Banking, Health Care, Content Outsourcing Services, Engineering Services Mumbai.

EffiQ e-Services Pvt. Ltd. Foreign Language Translations, Data Digitization, OCR Services, XML Tagging, e-publishing, Transaction Processing Pune.

EFunds Corporation Accounting, Contact Center /

Customer Care, Transaction Processing Chennai, Mumbai, Gurgaon Listed Company: NYSE 6,000 Fortune 500 and FTSE 100 companies.

eggFirst advertising design Offshore Advertising Design Mumbai.

Einsoft Solution Business Transcription Services, Inbound / Outbound voice, Contact Center / Customer Care, Data Conversion, Advanced Tech Support Chennai.

EKAM Infotech Pvt. Ltd. Accounting & Finance, Payroll, Tax Forms, Tax Management Gurgoan.

EKM Infotech Pvt. Ltd. Accounting & Finance, Accounts Receivable / Payable / Reconciliation Gurgaon.

Elico Ltd Medical Coding Services, Revenue Cycle Management Services, Healthcare Claims Processing, Healthcare Payer, Process Consulting, Medical Billing Hyderabad.

Elsoft Technologies Pvt. Ltd. Form Processing, Data Capture & Processing, Medical Transcriptions, OCR Services, XML Tagging, Medical Billing Bangalore.

eMR Technology Ventures (P) Ltd. Claims Processing, Data Capture & Processing, Credit Card Processing, Accounts Receivable / Payable / Reconciliation, Healthcare Claims Processing, Adjudication of Claims Gurgaon.

Enablizer Backoffice Technologies (P) Ltd. Accounting & Finance, Tax Forms, Process Outsourcing, Tax Management, Process Consulting Noida.

EncoreIndia Data Processing / Management, Technical Support / IT Helpdesk, CRM, Insurance, Insurance Administration, Offshore Advertising Design Visakhapatnam.

End-End Software (Asia Pacific) Pte. Ltd. SLA Monitoring, Application Performance Monitoring Singapore.

Endeavour Info Consulting Services (P) Ltd. Data Processing / Management, Claims Processing, Contact Center / Customer Care, Benefit Administration Services (PF/ESIC), Credit Processing Services, Content Outsourcing Services Bangalore.

ePeople HR Services Pvt. Ltd. People Solutions, HR, Payroll, Benefit Administration Services (PF/ESIC), Staffing Solutions and Training Bangalore.

Ephinay (Formerly AcFin) Transaction Processing, Accounts Receivable / Payable / Reconciliation Gurgaon.

EpiSource Pvt. Ltd. Accounting & Finance, e-publishing, Medical Coding Services, Offshore Advisory Services, Engineering Services, Medical Billing Chennai.

Epitome Global Services Pvt. Ltd. Finance, Business Research For Banking Industry, Offshore Advisory Services, Process Outsourcing Mumbai.

EServitium Accounting, eLearning, e-publishing, Chat and Email Support, Data Services Ahmedabad.

Espire Infolabs Data Capture & Processing, Medical Transcriptions, Inbound / Outbound voice, Data Conversion, Finance, Telecom Sector, Helpdesk Services New Delhi, USA, UK, Australia.

ESS Solutions Data Processing / Management, CRM, Accounting & Finance, Medical Transcriptions, Inbound / Outbound voice, Health Care USA, UK and India.

EStreet Inc Inbound / Outbound voice, Contact Center / Customer Care, Credit Card Processing, Data Conversion, Data Entry, Loan/Mortgage Processing Chennai.

Euclid Solutions Pvt. Ltd. Data Processing / Management, Document Management, Data Conversion, Data Entry, e-publishing Mumbai.

Exceed Outsourcing Private Limited. People Solutions, Accounting & Finance, Data Entry, Accounts Receivable / Payable / Reconciliation, Payroll New Delhi.

ExcellenceTech (An Enterprise by Kariwala Industries Ltd.) Accounting & Finance, Medical Transcriptions, Transaction Processing, Accounts Receivable / Payable / Reconciliation, Payroll, Tax Forms Calcutta.

ExlService Holdings, Inc. Technical Support / IT Helpdesk, Accounting & Finance, Insurance, Banking, Collection Services, Process Consulting Noida (3), Pune.

Eyecircus Studio Pvt. Ltd. Animation Ahmedabad Subsidary Company of Eyecircus UK.

FarBPO, LLC Accounting & Finance, Billing, Banking, Health Care, Receivables Management Bangalore, New York Private 5 Equity research companies, investment banks

Ficussoft India Ltd. Data Capture & Processing, Data Conversion, Data Entry, XML Tagging Bhubaneswar.

Fin-e-ssential Infotech India Ltd. HR, Payroll, Benefit Administration Services (PF/ESIC), Tax Management Mumbai.

Flex Point Marketing Pvt. Ltd. Accounting & Finance, Inbound / Outbound voice, Data Entry, Banking, Business Research For Banking Industry Hyderabad.

Fomax Information Technologies Pvt. Ltd. eCRM Solutions, Animation, XML Tagging, e-publishing Bangalore.

Fortune Infotech Ltd. Form Processing, Data Capture & Processing, Document Management, Document Imaging, Healthcare Claims Processing Bangalore.

Future Tekh eLearning, Animation, Data Conversion, Data Digitization, e-publishing Chennai, UK.

Galagali Multimedia Pvt. Ltd. eLearning, Animation, Data Conversion, Data Digitization, e-publishing, Document Imaging

Mumbai.

Gandhi InfotTech Pvt. Ltd. Tele-Marketing, Lead Generation Delhi 100 Banks, Telecom Companies, Financial Institutions

Gateway Group of Companies Data Processing / Management, Technical Support / IT Helpdesk, Network and Security Management, XML Tagging, Process Outsourcing, Offshore Advertising Design Ahmedabad.

Gausa India Limited. eLearning, CRM, Supply Chain Services, Legal Outsourcing New Delhi, Maryland [USA].

GeBBS Healthcare Solutions Claims Processing, Medical Coding Services, Health Insurance, Healthcare Payer, Medical Billing Mumbai Private 350 Leading Providers (Hospitals and Medical Billing companies) including Health and Hospital Corporation, Misys etc. Leading Payers (TPAs and Health Plans) including Plexis, Northstar, WebTPA, etc.

Genius Digital Solutions Form Processing, Document Management, Data Conversion, XML Tagging, e-publishing, Litigation chennai.

Genuine Infotech Private Limited. Data Processing / Management, Accounting & Finance, Data Entry, Accounts Receivable / Payable / Reconciliation, Payroll, Legal Outsourcing Coimbatore Private Limited 25 N.A.

Geo Surveys Pvt Ltd. Data Conversion, Data Entry, Data Digitization, CAD/GIS Chennai.

Gleam Info Services Pvt. Ltd. Medical Transcriptions, Data Entry Bangalore.

Global Infovision Data Processing / Management, Claims Processing, Insurance Claims Processing, Data Conversion, Medical Billing Hyderabad.

Global vantEdge Tele-Marketing, Collection Services,

Receivables Management Gurgaon (2) ChrysCapital Globalitstaffing HR, Recruitment Services, Staffing Solutions and Training Bangalore, Visakhapatnam, Europe.

GLOPORE IMS Data Processing / Management, Technical Support / IT Helpdesk, Network and Security Management, Remote Network Monitoring, Helpdesk Services Bangalore.

Go4customer Tele-Marketing, Inbound / Outbound voice, Market Research, Collection Services, Inbound Call Center Jaipur.

Graphinet Media Content Management, eCRM Solutions, Animation, Offshore Advertising Design Pune.

Great9 Info services Private Limited . Tele-Marketing, HR, Inbound / Outbound voice, Knowledge Management, Process Consulting Hyderabad.

Greenpoint Services Loan/Mortgage Processing Bangalore.

GSS America Inc. Form Processing, Data Capture & Processing, Data Conversion, XML Tagging, e-publishing, Litigation Hyderabad, Chicago.

GTL Ltd. Technical Support / IT Helpdesk, Customer Management, Customer Acquisition Mumbai GTL Limited, Internal accruals 950 Fortune 500 clients in Telecom, Insurance, Credit cards and Healthcare

Hartron Communications Ltd. Claims Processing, Transaction Processing, Health Insurance, Medical Billing Gurgaon.

HCL Infinet Ltd. Technical Support / IT Helpdesk, CRM, eCRM Solutions, Contact Center / Customer Care, Network and Security Management, Interactive Voice Response Mumbai.

HCL Technologies BPO Services Accounting, HR, Contact Center / Customer Care, Transaction Processing, Finance, Payroll, Admin Bangalore, Chennai, Noida [Delhi], Belfast [UK],

Malaysia - British Telecom.

Health Office (India) Pvt. Ltd. Medical Transcriptions, Medical Coding Services, Medical Billing Bangalore.

Heartland Medical Transcriptions Bangalore, Coimbatore, Delhi – 2000.

HeroITES Tele-Marketing, Contact Center / Customer Care, Customer Acquisition, Lead Generation, Data Validation Services Gurgaon A division of The Hero Group 1000 Fortune 500 client in US, UK and Australia.

Hewitt Outsourcing Services (India) HR Bangalore, Hyderabad, Chennai, Pune, Mumbai, Delhi A joint venture of View India Enterprises (VIE) and Hewitt Associates 210 Philips, IBM, ICI, Pfizer, BASF, Nokia, MindTree, Bank Muscat

Hi Q Process Transaction Processing, Travel, Transportation & Hospitality Industry Mumbai, Pittsburgh Private Airlines, GDS / CRS Service Providers, Consolidator, Travel Agents Hotels, Cruise-lines, Travel Portals Car Rentals, Trucking, Courier companies

Hinduja TMT Ltd. [HTMT] Technical Support / IT Helpdesk, Tele-Marketing, Insurance Claims Processing, Insurance, Telecom Sector, Health Care, Inbound Call Center Bangalore, Manila.

Hitech Infosoft Accounting & Finance, Document Management, Animation, Healthcare Claims Processing, Medical Billing Ahmedabad.

HLA Information Technologies Pvt Ltd Data Processing / Management, Data Capture & Processing, Data Conversion, Document Imaging Chennai Sudarsan Group of Companies 100 Triveni Engineering Management & Services [Chennai],SIS Technologies [New Delhi].

HnS Infotech Content Management, eLearning, Data Conversion, Data Entry, XML Tagging, Litigation Chennai

Subsidiary of Hard n Soft Technologies Pvt. Limited (www.hnsonline.com) 125 Ashok Leyland,Caterpillar,Ford India Ltd, Hyundai Motor India Ltd, KELTRON, Kerala Government, Tamil Nadu Tourism Development Corporation, Income Tax Dept,Larsen & Toubro Ltd,Turbo Energy,Toyota Kirloskar Motor, Bangalore,TVS Motor Company,ELCOT, Accenture Ltd etc.,

HOV Services Limited. Accounting & Finance, Transaction Processing, Accounts Receivable / Payable / Reconciliation, Insurance, Healthcare Claims Processing, Tax Management Pune, USA * 815 Some of the largest US based companies in the telecommunications, healthcare, insurance & finance and banking Industries

HTBPO e-Services Form Processing, Data Capture & Processing, Data Conversion, Data Entry, e-publishing Vellore.

Hughes BPO Services Technical Support / IT Helpdesk, Customer Interaction Management Gurgaon A division of Hughes Software Systems.

Husys Consulting Pvt. Ltd. HR, Contact Center / Customer Care, Employer Services, Staffing Solutions and Training Hyderabad.

i Consultancy Services Inbound / Outbound voice, B2B Lead Generation, Contact Center / Customer Care, Credit Card Processing, Inbound Call Center Kolkata.

I- Adore Services Tele-Marketing, Business Transcription Services, B2B Lead Generation,f Contact Center / Customer Care, Collection Services, Business Services Provision Noida.

i-serve Systems Pvt. Ltd. Accounting, Technical Support / IT Helpdesk, Tele-Marketing, Inbound / Outbound voice, Contact Center / Customer Care, Airfare Filling & Pricing Support Gandhinagar, Gujarat.

i-Spectra Solutions Claims Processing, Data Conversion, Data Entry, Litigation, Medical Coding Services, Medical Billing

Chennai .

i2c World Contact Center / Customer Care Mumbai.

I2iEnterprise Ltd. HR, Inbound / Outbound voice, Payroll Mumbai, Delhi, Chennai, Bangalore, Kolkatta, Chandigarh, Hyderabad, Ludhiana, Amritsar, Kanpur, Jaipur, Varanasi, Patna, Bhopal, Indore, Baroda, Surat, Rajkot, Pune, Nagpur, Vizag, Coimbatore, Madurai, Cochin & Trivandrum Private 1800 Large Indian / US based companies.

IBH Publishing Services Content Management, Data Capture & Processing, e-publishing, Finance, Fulfillment Services, Document Imaging Mumbai, New Jersey, New York.

IBN Technologies Ltd. Data Processing / Management, Inbound / Outbound voice, Contact Center / Customer Care, Finance, Lead Generation, Chat and Email Support Pune.

IBT India Ltd. Data Processing / Management, Data Conversion, Data Entry, Data Digitization, e-publishing, Data Mining Kochi 1500 UK, USA based companies.

iCall India Ltd. Tele-Marketing, Accounting & Finance, Contact Center / Customer Care, Knowledge Management, Corporate Secretarial Services, Loan/Mortgage Processing Ahmedabad.

ICallBiz Tele-Marketing, Inbound / Outbound voice, Contact Center / Customer Care, Loan/Mortgage Processing, Chat and Email Support, Data Services Ahmedabad.

ICICI oneSource Contact Center / Customer Care, Transaction Processing, Finance, Telecom Sector Bangalore, Mumbai $ 40 Million from ICICI Bank 4000 Fortune 500 and FTSE 100 companies, Virgin Atlantic iEnergizer, Inc HR, Accounting & Finance, Inbound / Outbound voice, Credit Card Processing, Banking, Collection Services, Payroll Noida.

iFactor Multimedia Technologies Pvt. Ltd eLearning, Animation, e-publishing Bangalore.

IFocus Business Solutions Limited Liability Co. Process Solutions, Inbound / Outbound voice, B2B Lead Generation, Knowledge Management, Loan/Mortgage Processing, Legal Research NCR.

IGate Technical Support / IT Helpdesk, Contact Center / Customer Care Bangalore Listed Company: NASDAQ GreenPoint Mortgage & ING Vysya IMAA Offshore Advertising Design Bangalore A division of MAA Group IMAX Solutions Data Processing / Management, Data Capture & Processing, Data Conversion, Data Entry, e-publishing, Document Imaging Chennai.

Imercius Technologies Pvt. Ltd. Technical Support / IT Helpdesk, Inbound / Outbound voice, Knowledge Management, Advanced Tech Support Mumbai Imercius Technologies is a BPO venture in India promoted by MIRC Electronics, a US$ 250 Million group that owns the brand ONIDA, - a household name 150 US and UK based clients

Immaculate Intractions India Limited. Technical Support / IT Helpdesk, Document Management, Inbound / Outbound voice, Surveys, Collection Services, Chat and Email Support Bangalore.

InBPO Limited Claims Processing, Inbound / Outbound voice, Accounts Receivable / Payable / Reconciliation, Chat and Email Support, Medical Billing Pune.

Indecomm Global Services Accounting & Finance, Document Management, Banking, Health Insurance, Loan/ Mortgage Processing Bangalore.

Indoserve Data Processing / Management, Tele-Marketing, Accounting & Finance, Accounts Receivable / Payable / Reconciliation, Inbound Call Center Noida.

Indus Markfin Pvt. Ltd. Inbound / Outbound voice, Contact Center / Customer Care, Collection Services Mumbai.

Indyasoft Claims Processing, Medical Transcriptions, Medical Coding Services, Medical Billing Delhi,Gurgaon.

Ineyva Technologies Form Processing, Data Processing / Management, Data Conversion, Data Entry, Litigation, Medical Billing Trichy.

Wfnfiniteoutsourcing Data Capture & Processing, Chat Support, CRM, Banking, Chat and Email Support, Engineering Services Gurgaon.

Infinity Ventures (I) Pvt. Ltd. Technical Support / IT Helpdesk, Remote Network Monitoring, Corporate Secretarial Services, Chat and Email Support, Advanced Tech Support, Technical Writing Noida.

Infoscript Services Pvt. Ltd. Medical Transcriptions Bangalore.

InfoSearch BPO Services Pvt. Ltd. Inbound / Outbound voice Chennai (2) InfoSearch Solutions Pvt. Ltd. 100 UK outbound - Telecom, USA - Outbound - Services, BPO - Italian Holiday rental company, Outbound - UK - Utility services.

Infotronics Pvt. Ltd. Tele-Marketing, Data Capture & Processing, Inbound / Outbound voice, B2B Lead Generation, Contact Center / Customer Care, Surveys Chennai.

InnovaQ Technologies Technical Support / IT Helpdesk, Chat Support, HR, Inbound / Outbound voice, Presales Chennai, UK, US.

Inox Global Services Limited. Data Capture & Processing, Accounting & Finance, B2B Lead Generation, Contact Center / Customer Care, Transaction Processing Gurgaon Part of Inox Group 100 Major clients from Americas, Europe and Australia

Inspire Solutions Private Limited. Form Processing, Data Processing / Management, Email Support, Document Management, Data Mining, Data Services Chennai Private Limited Company 18 Beverages Company Financial Services

Company.

Intec Infocom Pvt. Ltd. CAD/GIS Gurgoan Private Organisation 90 Canada, US, Europe

Integrated Language Solutions Foreign Language Translations New Delhi.

Integrated Software Solutions Pvt. Ltd. Email Support, Insurance, Finance, Telecom Sector, Health Care, Automotive Hyderabad.

Integreon Content Management, eLearning, Document Management, Knowledge Management, Business Research For Banking Industry, Legal Outsourcing London, Los Angeles, Mumbai, New York.

Intelenet Global Services Ltd. Technical Support / IT Helpdesk, Accounting & Finance, Contact Center / Customer Care, Transaction Processing, Insurance, Banking, Airline Chennai, Mumbai, US, UK HDFC, Barclays Bank plc. 4500.

Intelli-Corp India Business Solutions Technical Support / IT Helpdesk, Process Solutions, Network and Security Management, Remote Network Monitoring, Statutory Compliance, Process Consulting Hyderabad, Gurgaon.

Intelligent Computer Technology Form Processing, Data Processing / Management, Data Conversion, Data Entry, Data Digitization, e-publishing New Delhi.

Intellisight India Pvt. Ltd. Billing, Health Care, Health Insurance, Healthcare Claims Processing Hyderabad.

Inter World Contact Center Inbound / Outbound voice, Contact Center / Customer Care, Lead Generation Bangalore.

Interactive Web Resources Content Management, Chat Support, Data Entry, Admin, Chat and Email Support, Content Outsourcing Services Delhi.

Intercorridor Systems Ltd. Technical Support / IT Helpdesk,

Inbound Call Center, Advanced Tech Support, Technical Writing Mumbai.

InterWorld Contact Center Inbound / Outbound voice Bangalore A division of Veetech Infoline Pvt. Ltd.

inTouch Analytics Pvt. Ltd. CRM, Customer Acquisition, Data Mining, Customer Profiling, Market Analytics Bangalore 30 Leading names in the Retail and CPG sectors in India

InTouch Solutions Tele-Marketing, Inbound / Outbound voice, Tax Forms, Customer Interaction Management, Legal Outsourcing, Logistics Tracking New Delhi.

Invitra Technologies Content Management, Data Processing / Management, eCRM Solutions, Animation Pune.

iRadix technologies. Form Processing, Data Conversion, Data Entry, Data Digitization, XML Tagging, e-publishing Chennai, Pondicherry, Trinelveli, Hong Kong Privately funded 90 Confidential

Iridium BPO Division (Geosoft Technologies Trivandrum Limited) Medical Transcriptions, Medical Coding Services Trivandrum.

ISeva, Inc. Loan/Mortgage Processing, Advanced Tech Support Bangalore (3), Chennai (1)

e4e Labs [$8m in Feb 2002] 1500 Fortune 500 companies in US and UK

ISN Data Processing Pvt. Ltd. Form Processing, Data Processing / Management, Data Capture & Processing, Data Conversion, Data Entry, OCR Services, Remittance & Payment Processing, Check Processing.

IT Global Solutions Inc Network and Security Management, Remote Network Monitoring, Advanced Tech Support New Delhi.

ITCube Solutions Pvt. Ltd. Accounting & Finance, Inbound / Outbound voice, Market Research Pune.

ITech Workshop Form Processing, Claims Processing, Insurance Claims Processing, Health Care, Healthcare Claims Processing, Medical Billing Bangalore - 15 U.S. based Billing Service Bureaus, Healthcare Solution Providers, Clearinghouses, Hospitals & Clinics and Indian Healthcare BPOs

Jadian Technologies Data Processing / Management, Data Conversion, Data Entry, Data Services Chennai.

JAY AR Solutions Form Processing, Data Processing / Management, Data Capture & Processing, Inbound / Outbound voice, Data Conversion, Data Entry Chennai.

Jiva InfoTech Tele-Marketing, Inbound / Outbound voice, Contact Center / Customer Care, Customer Acquisition, Loan/ Mortgage Processing, Advanced Tech Support Hyderabad.

Kaizen Global Services Technical Support / IT Helpdesk, Tele-Marketing, Inbound / Outbound voice, Contact Center / Customer Care, Collection Services, Helpdesk Services Navi Mumbai.

KARVY Global Services Accounting, Market Research, Transaction Processing, Payroll, Adjudication of Claims, Market Analytics Hyderabad, New York, California, Toronto, Zurich and London.

KARVY Global Services Limited Accounting & Finance, Contact Center / Customer Care, Knowledge Management, Transaction Processing, Payroll, Health Care Hyderabad KARVY Global Services Limited is a fully owned company of the KARVY Consultants Limited. 300 KARVY services over 350 corporates and over 18 million investors.

Katalyst Technical Support / IT Helpdesk, Remote Network Monitoring, Helpdesk Services, Chat and Email Support, Advanced Tech Support New York, Navi Mumbai, Bangalore.

Katwa BPO Services Pvt. Ltd. Accounting & Finance,

Accounts Receivable / Payable / Reconciliation, Payroll, Tax Forms Belgaum.

Keane Worldzen India Private Limited. Insurance, Collection Services, Health Care, Financial Transcription Gurgaon A Keane Group Company. Listed Company: NYSE 400 Leading U.S. and U.K. companies

Kilburn Software Technologies India Limited. Document Management, Data Conversion, Data Digitization, Document Imaging Kolkata (Calcutta) Funded by Kilburn Group 30 Ministry of Education, United Bank of India, State Bank of India, Economic Times, FDC Limited, Strassenburg Pharmaceuticals, Telerama India Ltd, Himachal Telematics, Himachal Futuristic Communications Ltd, India Foils Ltd., Light Metal Industries Ltd., Cans & Closures Ltd, Siemens (India) Ltd.

KK Technologies Data Conversion, Data Entry, XML Tagging, e-publishing, Data Services Vijayawada.

KNM Services (P) Ltd. Form Processing, Data Entry New Delhi.

Knoah Solutions Contact Center / Customer Care, Inbound Call Center, Chat and Email Support Hyderabad.

Kosher Infotech Data Processing / Management, Data Conversion, Data Entry, Payroll Bangalore.

KPO Guru Content Management, Knowledge Management, Animation, Legal Research Ahmedabad, Kochin.

KPO Web Outsourcing Services Data Processing / Management, Data Capture & Processing, Accounting & Finance, Data Entry, Data Digitization, Data Mining New Delhi.

Kraftwagen Design Pvt. Ltd. Accounting & Finance, Market Research, Contact Center / Customer Care, Insurance, Business Research For Banking Industry, Airline Pune 100%.

Krd Infotech Private Limited Form Processing, Data

Capture & Processing, Accounting & Finance, Data Conversion, Data Entry, e-publishing New Delhi.

Kshema BPM Content Management, Claims Processing, Invoice Processing, Health Insurance, Loan/Mortgage Processing Bangalore.

Lancesoft India Pvt. Ltd. Data Processing / Management, Medical Transcriptions, Payroll, Medical Coding Services, Medical Billing Chennai.

Lapiz Digital Services Litigation, Legal Outsourcing Chennai Public Limited Company 360 US and UK.

Last Peak Pvt. Ltd. Medical Transcriptions, Medical Coding Services, Healthcare Claims Processing, Medical Billing Calcutta.

Lawkim India Pvt. Limited. Technical Support / IT Helpdesk, Telecom Sector, Manufacturing, Customer Interaction Management Mumbai.

Lemurian Technologies Form Processing, Claims Processing, Data Capture & Processing, e-publishing Chennai.

Lester Incorporated Content Management, Data Capture & Processing, Chat Support, Data Entry, Accounts Receivable / Payable / Reconciliation Connecticut, Mumbai.

Litmuz Business Transcription Services, Accounting & Finance, Contact Center / Customer Care, Data Entry, XML Tagging, e-publishing Bangalore.

Live Help BPO Technical Support / IT Helpdesk, Inbound / Outbound voice, Contact Center / Customer Care, Customer Interaction Management, Chat and Email Support Hyderabad, Vizag, Guntur Division of Essel Software & Services Ltd.

LNSEL Inbound / Outbound voice, Accounts Receivable / Payable / Reconciliation, Collection Services, Chat and Email Support, Data Services Kolkata.

LogicaCMG pvt limited. eLearning, Claims Processing, HR, Accounting & Finance, Contact Center / Customer Care, Data Services Bangalore , Czech , UK , Netherlands.

Ma Foi Outsourcing Solutions Limited. HR, Payroll, Resume Management Services Headquarters: Chennai.Other locations are: Bangalore, Mumbai,New Delhi, Kolkata, Pune, Ahmedabad, and Hyderabad.

Overseas Locations are: Dubai, London and Singapore - 100 Servicing top corporates

Macmillan Information Processing Services (MIPS) eLearning, Data Conversion, e-publishing, Fulfillment Services, Content Outsourcing Services, Offshore Advertising Design Bangalore,Chennai,USA and UK A Division Of Macmillan India Ltd.Listed Company: Bombay Stock Exchange, National Stock Exchange 1700 Leading Publishing Companies, Universities and Associations.

Macro Infotech Pvt. Ltd. Form Processing, Data Capture & Processing, Data Conversion, Data Entry, OCR Services, Document Imaging Bangalore.

MacroTrans Medical Transcriptions, Medical Coding Services, Medical Billing Hyderabad Macrotrans Inc.

Magnum Multimedias Form Processing, Data Processing / Management, Data Capture & Processing, Email Support, Chat Support, Data Entry Madurai.

Mahima Technology Pvt Ltd Claims Processing, Data Capture & Processing, Data Conversion, Data Entry, Data Digitization Salem.

Mainstay Teleservices (P) Ltd. Inbound Sales, Email Support, Chat Support, Contact Center / Customer Care, Lead Generation, Data Validation Services, Tax Management Bangalore Tecapet and Comcraft [UK based firms] 120 US, UK, Canada based companies.

MANAM Consultancy Services PVT Ltd Form Processing, Data Capture & Processing, Data Conversion, XML Tagging, e-publishing, Litigation Chennai, USA, Australia Part of the MANAM Group 250 Australia, UK, USA Publishers

Manisha Infocom Services Pvt. Ltd. Form Processing, Data Capture & Processing, Transaction Processing Aurangabad.

Manjushree Infotech Data Capture & Processing, Insurance Claims Processing, Medical Transcriptions, Inbound / Outbound voice, Data Conversion, Transaction Processing, Remote Network Monitoring, Medical Billing Kolkata, Bangalore.

Man oeuvre Infotech Email Support, Knowledge Management, Data Entry, Health Care, Business Research For Banking Industry, Healthcare Claims Processing kolkata.

Mantas Group Document Management, Animation, Data Conversion, Data Entry Delhi.

Manthan Services Data Mining, Legal Outsourcing, Retail Customer Support Bangalore.

Mark World Content Management, Insurance Claims Processing, Knowledge Management, Banking, Loan/Mortgage Processing, Process Outsourcing New Delhi.

Markelytics International Data Processing / Management, Market Research, Surveys, Collection Services, Data Validation Services, Data Services Bangalore Private Funding 40 Fortune 500 companies from USA and Europe.

Marvel BPO Technical Support / IT Helpdesk, Tele-Marketing, CRM, Contact Center / Customer Care, Lead Generation, Loan/Mortgage Processing Gurgaon Internal 40 Assent Europe, vServe Australia.

Med-Dox Solutions Form Processing, Insurance Claims Processing, Medical Transcriptions New Delhi.

Med global services Insurance Claims Processing, Medical Transcriptions, Health Care, Medical Coding Services, Medical Billing Noida.

Medicare's Pvt. Ltd. Medical Transcriptions Pune Private 25 Multiple MTSOs in the US.

MedilandTranstech Pvt. Ltd. Accounting & Finance, Medical Transcriptions, Medical Coding Services, Medical Billing Hyderabad, New Jersey Private Limited 180 Hospitals, Group Practices, Clinics etc in USA

Med soft India Pvt. Ltd. Medical Transcriptions, Medical Coding Services Bangalore, Mysore, Coimbatore, Sunnyvale (USA) Privately funded 270 US Based Health Care Service Providers.

Medusind Solutions Medical Transcriptions, Medical Coding Services, Revenue Cycle Management Services, Healthcare Claims Processing, Member Management, Medical Billing Chennai, Mumbai.

Mercury Communication Content Management, Content Outsourcing Services Hyderabad.

Mercury Outsourcing Management Limited Tele-Marketing, Insurance Claims Processing, Market Research, Lead Generation, Loan/Mortgage Processing Secunderbad.

Micropro Software Solutions Pvt. Ltd. Content Management, Data Processing / Management, Network and Security Management, Health Care, Remote Network Monitoring, Advanced Tech Support Nagpur.

Minds eye e-publishing, Offshore Advertising Design Mumbai, New York, Washington DC.

Modular InfoTech (P) Ltd Content Management, Data Conversion, Data Entry, Data Digitization, XML Tagging, e-publishing Pune Private Limted 180 • Microsoft Inc. USA, • Sun Microsystems, USA • Excellead Technologies, Japan • DX

Technologies, Singapore • Hindustan Computers Ltd. • TVS Electronics Ltd. • WeP Peripherals Ltd. • HP, India Ltd. • WEBEL • CMC • Wockhardt Ltd. • Lipi Data systems • Telco Ltd. • Rediffmail Ltd. • Kaveri Communications Ltd. • Deccan Herald News Papers • Sakal Papers Limited • Pune Municipal Corporation.

Molecular Connections Pvt. Ltd. Data Mining Bangalore Healthcare business incubator 30 Biotechnology, Pharmaceutical, Genomics / Proteomics, Bioinfomratics, Target Validation Evaluation

Motif India Accounting, Insurance Claims Processing, Email Support, Chat Support, Order Processing, Payroll, Credit Appraisals, Fulfillment Services Ahmedabad –

Move Offshore Call Center Tele-Marketing, Inbound / Outbound voice, Data Entry, Customer Acquisition, Lead Generation Chennai.

MphasiS BPO Content Management, Accounting & Finance, Credit Card Processing, Banking, Airline Ahmedabad, Bangalore, Mangalore, Mumbai, Mexico, Noida, Pune, Shanghai, Singapore, UK, Australia.

MPower Digitech Ltd. Form Processing, Data Processing / Management, Technical Support / IT Helpdesk, Contact Center / Customer Care, Chat and Email Support Chennai Internal 63 USA, UK, Australia.

mRecord Healthcare Solutions Medical Transcriptions, Medical Billing Mumbai - 100 US-based Medical Practices and Hospitals.

MTportal Medical Transcriptions, Accounts Receivable / Payable / Reconciliation, Medical Coding Services, Medical Billing Noida.

Munimjee Accounting & Finance, Accounts Receivable / Payable / Reconciliation, Benefit Administration Services (PF/

ESIC), Process Outsourcing, Tax Management Ahmedabad.

Mysoft International Pvt. Ltd. Form Processing, Claims Processing, Medical Transcriptions, Contact Center / Customer Care, Data Entry, Medical Billing Bangalore.

Nash Infotech Accounting, Form Processing, Data Processing / Management, Chat Support, Inbound / Outbound voice, Credit Card Processing, Animation, Data Conversion, OCR Services, XML Tagging, Check Processing Nasik.

Nebula Consulting B2B Lead Generation, Resume Management Services, Technical Writing Bangalore.

NECTA TECHNOLOGY (P) Limited. Technical Support / IT Helpdesk, Inbound / Outbound voice, B2B Lead Generation, Contact Center / Customer Care, Lead Generation, Loan/ Mortgage Processing Chennai.

Net Connect India Email Support, Data Conversion, Data Entry, Health Insurance, Process Outsourcing, Medical Billing Ahmednagar - 35 Worldwide.

Netcreativemind Solutions Pvt. Ltd. Data Processing / Management, Contact Center / Customer Care, Customer Management, Helpdesk Services, Chat and Email Support, Advanced Tech Support New Delhi.

NetEdge Computing Solutions Pvt. Ltd. eLearning, Data Processing / Management, Technical Support / IT Helpdesk New Delhi, Noida, Bangkok, London, New York

NetPolaris Inc Medical Transcriptions, Medical Coding Services, Healthcare Claims Processing, Medical Billing Kolkata, IN / MA, USA.

Netscribes India Limited Market Research, Knowledge Management, Business Research For Banking Industry, Sales Intelligence, Competitor Analysis, Content Outsourcing Services Mumbai.

Netspace Communications Pvt. Ltd. Data Processing / Management, Tele-Marketing, Inbound Sales, B2B Lead Generation, Contact Center / Customer Care, Helpdesk Services Delhi.

Net Sutra Data Processing / Management, Technical Support / IT Helpdesk, Document Management, Data Conversion, e-publishing, Chat and Email Support Delhi.

Neural IT Data Processing / Management, Claims Processing, Document Management, Data Mining, Document Imaging, Legal Outsourcing Mumbai.

Nextwave Multimedia Content Management, eLearning, Animation Chennai Not funded 35 Kris Srikkanth, Ashok Leyland, Apollo, Sify, Polaris, Saint Gobain etc.

Nexustronics Datasystems Form Processing, Claims Processing, Data Capture & Processing, Data Conversion, Data Entry, Check Processing Mumbai.

Niche Marker Callnet Pvt. Ltd. Tele-Marketing, Inbound Sales, Inbound / Outbound voice, B2B Lead Generation, Surveys, Health Care, Inbound Call Center New Delhi.

NIIT Smart Serve Ltd. Technical Support / IT Helpdesk, Inbound / Outbound voice, Collection Services, Insurance Administration, Statutory Compliance Gurgaon.

NIIT Technologies Ltd. 740 Sesame Ltd (Misys Plc), Telecom providers in UK, Direct marketing agencies

Nipuna Services Ltd. Claims Processing, Data Capture & Processing, HR, Accounting & Finance, Credit Card Processing, Animation, Customer Interaction Management, Loan/Mortgage Processing Bangalore, Hyderabad Subsidiary of listed Company: Satyam Computers, $20m from Intel Capital and Olympus Capital [Oct 2003] 1100 Elsevier,Large Scientific Publishing House based in the UK, US Telecom Service Provider, US Large Insurance Company, US Large Engine

Manufacturer, US Large Healthcare Company.

Nirvana Business Solutions Technical Support / IT Helpdesk, Accounting & Finance, Contact Center / Customer Care, Transaction Processing, Data Mining, Marketing Program Support Ban.

Niyati Technologies Animation, Content Outsourcing Services Chennai No 12 InteQ Inc, Janeeva Inc, Vox Spectrum Inc, Mindspring Inc, ZMapping, Cybernet Software Systems, Ready Test Go, Benchmark Softech, STC Technologies, Bahwan Cybertek, IBS Consulting Group, ReviewNet Corporation, SolutionNet, Stringinfo, Timna Telecoms.

Oasis Mediscript Pvt. Ltd. Claims Processing, Insurance Claims Processing, Medical Transcriptions, Medical Coding Services, Claims Administration, Medical Billing New Delhi.

Ocwen Financial Solutions Private Ltd. Accounting, Tele-Marketing, Contact Center / Customer Care, Collection Services, Payment Services, Insurance Administration, Agency Management, General Administration, Residential & Commercial Loan Servicing Bangalore, Mumbai.

Office Tiger Data Processing / Management, Accounting & Finance, Data Digitization, Business Research For Banking Industry New York, Atlanta, London, Frankfurt, Dallas, Salt Lake City, Chennai, Colombo, satellite operations and multiple client-site delivery centers $52 million [Francisco Partners] in June 2004. 3500 Fortune 500 companies and top retail chains, Big-4 accountancies, management consultancies, global law firms

Offshore Coding International, Inc. Data Processing / Management, Data Capture & Processing, Document Management, Data Conversion, Data Entry Bangalore.

OKS Group Inbound Sales, Data Capture & Processing, Email Support, Chat Support, Contact Center / Customer Care, Inbound Call Center Chennai, New Dehli, Noida, Canada, and

US Privately owned company 2000 Fortune 1000; Business Information companies, Telco's, IT, and Healthcare.

OKS SpanTech (P) Ltd. Data Processing / Management New Delhi.

Olympus InfoTech Pvt. Ltd. Foreign Language Translations, Data Capture & Processing, Data Conversion, Data Digitization, XML Tagging, e-publishing Chennai, Illinois (USA).

OMCAD SYSTEMS INDIA INC Data Conversion, Data Digitization, CAD/GIS Bangalore.

Omega Health IT Solutions Pvt Ltd Data Processing / Management, Claims Processing, Medical Coding Services, Healthcare Claims Processing, Medical Billing Hyderabad 35 US and Europian Clients

OP Universal Consultants Limited (OPUC) Technical Support / IT Helpdesk, Tele-Marketing, Inbound Sales, Inbound / Outbound voice, B2B Lead Generation, Inbound Call Center Noida.

Oremus Corporate Services Business Feasibility, Accounting & Finance, Statutory Compliance, Project Appraisal & Funding Hyderabad - India and US based companies

Orion Outsourcing Pvt. Ltd. Data Processing / Management, Insurance Claims Processing, Accounting & Finance, Credit Card Processing Mumbai Funded By Directors 50 India Tourism Development Corporation (ITDC), New India Assurance Company Limited, ICICI Bank Ltd,Punjab National Bank, State Bank of India, ACC limited,Textile Corporation of Marathwada Limited, United Phosphorous Limited, Synovate India Pvt. Ltd., Videocon International Limited, Wockhardt Limited, Zenith Limited.

Orion soft Technologies Form Processing, Email Support, Accounting & Finance, Data Conversion, Health Care, Medical Billing Chennai.

OS2i Pvt. Ltd. HR, Resume Management Services, Recruitment Services, Staffing Solutions and Training Pune International Investors 100 UK, USA, Canada, Middleeast Recruitment Agencies, Corporate Recuitment Functions.

OSIRIS Solutions Pvt. Ltd. Content Management, Inbound / Outbound voice, B2B Lead Generation, Market Research, Surveys, Chat and Email Support Chicago, Ahmedabad Privately Owned 185 Fortune 500 US Telecom, Banks.

Osource (India) P. Ltd. HR, Accounting & Finance, Accounts Receivable / Payable / Reconciliation, Payroll HQ & Back Office in Mumbai, Branches in Bangalore, Delhi & Pune.

Outsource Partners International, Inc. Accounting & Finance, Transaction Processing, Tax Forms, Document Imaging Bangalore, Kochi, Thiruvananthapuram - 450 Midsized and Large Enterprises, CPA firms.

Outsourzing Worldwide Pvt. Ltd. Accounting, Payroll, Periodical returns (GST,VAT) reporting, Management reporting Chennai.

Ozone Technologies Claims Processing, Data Capture & Processing, Litigation Chennai.

Pan Business Lists Pvt Ltd Form Processing, Data Capture & Processing, Process Solutions, Accounting & Finance, Document Management, Surveys, Knowledge Management, Data Conversion, Data Digitization, Data Mining, Data Validation Services, Process Outsourcing New Delhi.

Paramount Outsourcing Data Capture & Processing, Accounting & Finance, Inbound / Outbound voice, B2B Lead Generation, Telecom Sector, Chat and Email Support Chicago, Delhi, Nagpur, Amravati, Pune Computech services pvt. ltd. 170 Motorola PCS, Motorola CGISS, Delhi Jal board, Water boards in Maharashtra, Municipal Corporations, Municipal Councils, Wavelet group, Momentum Data System.

Parth Softech Tele-Marketing, Accounting & Finance, Inbound / Outbound voice, Loan/Mortgage Processing, Legal Outsourcing, Legal Research Noida.

Patni Computer Systems Industries Ltd. Technical Support / IT Helpdesk, Network and Security Management, Third Party QC/QA Chennai By Patni Computer Systems Technology Limited 1000 Banks, Engineering companies, Airlines.

Patwa Group Email Support, Chat Support, Accounting & Finance, Data Entry Mumbai.

Phoenix Global Solutions Technical Support / IT Helpdesk, Tele-Marketing, Contact Center / Customer Care Bangalore Wholly owned subsidiary of The Phoenix Companies Inc. (NYSE:PNX) Insurance, Banking, Finance, Consulting and Technology industries.

PINNACLE Technologies Private Limited. Accounting, HR, Inbound / Outbound voice, Contact Center / Customer Care, Transaction Processing, Finance, Payroll, Admin Pune-India, Romania-Europe and Illions- USA Private 47 UK, Continental Europe and USA based blue chip Companies in the Public and Private Sector.

PIT Solutions Email Support, Chat Support, Data Entry, Chat and Email Support, Content Outsourcing Services Trivandrum.

Planet soft Inc. Collection Services, Health Insurance Hyderabad India.

Pointsoft (P) Ltd. Surveys, Data Conversion, Data Digitization, Staffing Solutions and Training, CAD/GIS Hyderabad Private Limited 200 Utilities (Electric, Telephone, Water), NRSA, Municipalities & Local Government, Leading Engineering companies. Most of the clients from USA, Canada and Europe

Pollinex.net Accounting, Data Conversion, Data Entry, Data Digitization, OCR Services, Chat and Email Support Sangli.

Porteck India Info services Pvt. Ltd. Medical Transcriptions, Medical Coding Services, Healthcare Claims Processing, Medical Billing Noida Privately Funded 165 Doctors (group & solo), Medical Billing Companies, Surgical Centeres, Labs (radiology & pathology, Medical Transcription Companies, Hospitals, Nursing Homes, Old Age Care Homes.

PPTS IND PVT.LTD. Business Transcription Services, Medical Transcriptions, Data Mining, Financial Transcription Coimbatore.

Pradot Technologies Private Limited. Medical Transcriptions, Accounts Receivable / Payable / Reconciliation, Medical Billing, Financial Transcription Bangalore.

Prestige Health Administrator (P) Limited. Medical Coding Services, Health Insurance, Healthcare Claims Processing, Third Party QC/QA, Healthcare Payer, Medical Billing Bangalore.

Prime Business Solutions Tele-Marketing, Inbound / Outbound voice, Contact Center / Customer Care, e-publishing, Lead Generation Ghaziabad (NCR).

Prime Technologies Insurance Claims Processing, Accounting & Finance, Data Entry, Payroll, Tax Management, Medical Billing Ahmedabad (Gujarat) India.

Pringle Software Technologies Pvt. Ltd. Data Processing / Management, Claims Processing, Medical Transcriptions, Credit Card Processing, XML Tagging, Data Validation Services, Credit Processing Services, Medical Billing Hyderabad Private 14 Amazon.Com, Factor 360 Inc., Oahe Health Management, LLC, and other financial, education, retail and health care service providers.

PrivoCorp Loan/Mortgage Processing Bangalore, New Jersey.

ProAdept Technical Support / IT Helpdesk, Tele-Marketing, Inbound / Outbound voice, Contact Center / Customer Care,

Transaction Processing, Chat and Email Support Dehradun - 475 IMSI United States, Nursing Resources Incorporated, EastWest Technologies

ProAssign Services Pvt. Ltd. Accounting, Accounting & Finance, Accounts Receivable / Payable / Reconciliation, Finance, Tax Forms Hyderabad, New Jersey Private Limited 10 CPAs and SME segment in USA

Process Global, Inc Content Management, Technical Support / IT Helpdesk, eCRM Solutions, Inbound Call Center, Offshore Advisory Services, Process Outsourcing Santa Clara, Bangalore.

Process-Max Solutions P. Ltd. Business Transcription Services, B2B Lead Generation, Contact Center / Customer Care, Data Entry, Data Digitization, Competitor Analysis Pune.

ProcessMind Services Pvt. Ltd. Process Solutions, Accounting & Finance, Inbound / Outbound voice, Insurance, Banking, Finance, Revenue Collection, Medical Billing Bangalore.

Prodat It Solutions (India) Private Limited. Accounting & Finance, Order Processing, Knowledge Management, Accounts Receivable / Payable / Reconciliation, Payroll, Tax Forms Bangalore.

Progeon Insurance Claims Processing, Email Support, CRM, Accounting & Finance, Banking, Health Care Bangalore,Jaipur, Pune, Czech Republic Citigroup Investments[$20m in 2002], Subsdiary of Infosys 5,293 GreenPoint Mortgage [$30m for 5 years starting 2002], Cisco, British Telecom, Fidelity Capital

Prognosys e Services Market Research, Knowledge Management, Data Mining, Lead Generation, Competitor Analysis, Customer Profiling New Delhi , Bangalore.

ProLease India Pvt. Ltd. HR, Accounting & Finance, Payroll, Benefit Administration Services (PF/ESIC), Recruitment

Services, Corporate Secretarial Services, Secure Data Processing, Staffing Solutions and Training Chennai, Bangalore, Hyderabad, Delhi, Mumbai.

Promantra Synergy Solutions Ltd. Medical Transcriptions, Accounts Receivable / Payable / Reconciliation, Health Care, Medical Coding Services, Medical Billing Hyderabad.

PROMYX Data Processing / Management, Accounting & Finance, Accounts Receivable / Payable / Reconciliation, Check Processing, Legal Outsourcing Bangalore, Cochin.

Provestment Services Ltd. Business Feasibility, Accounting & Finance, Billing, Accounts Receivable / Payable / Reconciliation, Billing Collections New Delhi Promoters contribution 35 UGS PLM Solutions, Flex Industries Ltd, Printotech Global Limited, Prakash Hospital.

Puma source Ltd Data Capture & Processing, Accounting & Finance, B2B Lead Generation, Data Validation Services, Legal Outsourcing London, Delhi, Punjab.

Purple Support Accounting, Insurance Claims Processing, Billing, Collection Services, Payroll, Data Mining, Lead Generation, Advanced Tech Support Mumbai.

Pyramid Consulting Inc Content Management, Data Capture & Processing, Document Management, e-publishing, Accounts Receivable / Payable / Reconciliation Gurgoan Private 330 GeCIS, Accenture, UPS, BellSouth, Georgia Pacific.

Q-Pro Outsourcing Pvt. Ltd. Accounting & Finance, Transaction Processing, Tax Management Bangalore.

QED Enabled Solutions Private Limited. B2B Lead Generation, Lead Generation, Inside Sales Pune.

QMTINDIA.com Medical Transcriptions Siliguri.

Quadrant Data Solutions Inc. Data Processing / Management, Inbound / Outbound voice, Contact Center /

Customer Care, Knowledge Management, Data Mining Chennai, Philadelphia.

Qualcrest Data Solutions Accounting & Finance, Accounts Receivable / Payable / Reconciliation, Payroll, Tax Forms Jaipur Private 200 References available on request.

QualHub Solutions P. Ltd. Content Management, Data Processing / Management, Data Capture & Processing, Data Conversion, Data Entry, Data Digitization Hyderabad Private 130 Worldwide.

Quality Time Data Services Pvt. Ltd. Data Capture & Processing, Medical Transcriptions, Data Conversion, Data Entry, Finance, Medical Billing Pune & Chennai.

Quick Data Services Form Processing, Data Processing / Management, Data Capture & Processing, Document Management, Data Entry, Accounts Receivable / Payable / Reconciliation Kolkata.

Quintant Services Pvt. Ltd. Accounting & Finance, Business Services Provision Bangalore A subsidiary of iGATE 300 Quscient Business Services Accounting & Finance, Tax Management Chennai 100

R M FINANCIAL SERVICES Transaction Processing, Accounts Receivable / Payable / Reconciliation, Remittance & Payment Processing, Invoice Processing, Loan/Mortgage Processing, Billing Collections Cochin.

R Systems Inc Contact Center / Customer Care, Medical Coding Services, Revenue Cycle Management Services, Healthcare Claims Processing, Billing Collections, Medical Billing Sacramento, CA; New Delhi.

R.R. InfoTech Limited. Data Processing / Management, Data Capture & Processing, Data Conversion, e-publishing, Litigation, Medical Billing Chennai, Dallas (TX, USA).

RedHorns Software Pvt. Ltd. Data Capture & Processing,

Network and Security Management Chandigarh.

Regency business Consulting Private Limited. Data Capture & Processing, Data Conversion, Data Entry, Data Digitization, XML Tagging, e-publishing Chennai (India), Detroit (US) Funded by investor based in Detroit(United States) 114 Renaissance Pte. Ltd. Everonn India, CAMS India.

Rehash Infotech Data Processing / Management, Data Conversion, Data Entry Mumbai.

Resurgent India Ltd. Accounting & Finance, Payroll, Loan/ Mortgage Processing, Process Outsourcing, Tax Management New Delhi Its a closely held Limited Liability company 50 Vmart.

RevenueXL Inc. Accounts Receivable / Payable / Reconciliation, Medical Coding Services, Revenue Cycle Management Services, Healthcare Claims Processing, Medical Billing Jaipur, San.

Revons IT Solutions Pvt. Ltd. Claims Processing, Data Capture & Processing, Accounting & Finance, Data Conversion, Data Entry, CAD/GIS Chennai.

Rhombus Technologies Email Support, Chat Support, B2B Lead Generation, Contact Center / Customer Care, Data Entry Mumbai.

RICO ITes Technical Support / IT Helpdesk, Inbound / Outbound voice, Data Conversion, Data Entry, Remote Network Monitoring, Helpdesk Services Mumbai Rico is self-funded and financially strong. We are profitable as a company and have strong cash reserves to provide financial backing for future growth and investments. 50 AllStream Asia RICO Softech Ltd. Email Support, Inbound / Outbound voice, Contact Center / Customer Care, Transaction Processing Gurgoan Forms a division of RICO Auto Industries Ltd., a $220 million global enterprise. 150 USA, UK and Canada based companies

Rightangle Technologies Private Limited. Process Solutions, Document Management, Knowledge Management, XML Tagging, Resume Management Services New Delhi.

Rinam Infosys eLearning, Animation, Data Conversion, Data Entry, Data Digitization, XML Tagging Kolkata.

Ring India InfoTech Pvt. Ltd. Tele-Marketing, B2B Lead Generation, Collection Services, Lead Generation, Loan/ Mortgage Processing, Process Outsourcing New-Delhi Own 100 US ,UK, and AUS corporations.

RNC-BPOI Form Processing, Claims Processing, Document Management, Data Conversion, XML Tagging, e-publishing Trivandrum.

RSB Systems Pvt. Ltd. Accounting, Data Processing / Management, Claims Processing, Document Imaging, Healthcare Claims Processing Gurgaon.

RSKTechnologies Data Processing / Management, Business Transcription Services, Medical Transcriptions, Data Conversion, Data Entry Bangalore.

RT Outsourcing Services Ltd. Technical Support / IT Helpdesk, eCRM Solutions, Helpdesk Services, Chat and Email Support New Delhi Funded by Prime VC funding organisations 600 HP, Sony Ericsson, NCR, Philips, Samsung.

S2S IT Solutions Pvt. Ltd. Data Processing / Management, Inbound / Outbound voice, Transportation, Logistics Tracking Bombay.

Saboo e-Solutions Pvt. Ltd Accounting & Finance, Accounts Receivable / Payable / Reconciliation, Payroll, Tax Forms Kolkata.

Saitech Data Systems Pvt. Ltd. Medical Transcriptions, Data Conversion, Data Entry, Inbound Call Center Pune.

Samdisha Software Research & Development Center Pvt.

Ltd. Data Conversion, e-publishing Delhi.

Samtrack Technology Services Pvt. Ltd. Claims Processing, Accounting & Finance, Medical Transcriptions, Payroll, Healthcare Claims Processing Bangalore, Chennai, Hyderabad , Madurai, London, Ann Arbor (USA).

Sand Martin Consultants Pvt. Ltd. Claims Processing, Accounting & Finance, Payroll, Royalty Processing, Retirement Planning New Delhi.

SatyaDev Solutions Data Conversion, Data Entry, Data Digitization Bhubaneswar.

Scandent Group Technical Support / IT Helpdesk, Contact Center / Customer Care, Chat and Email Support Bangalore, Chennai, Mumbai.

Scope e-Knowledge Center Pvt. Ltd. eLearning, Data Capture & Processing, Market Research, Data Mining, Competitor Analysis, Content Outsourcing Services Chennai Infinity Venture Fund 300 8 including Fortune 500 companies in publishing & STM area.

Seaview Support Systems Pvt. Ltd. Business Transcription Services, Medical Transcriptions, Legal Outsourcing Trivandrum.

Selectronic Equipment & Services Pvt.Ltd. Medical Transcriptions, Data Conversion, Data Entry Gurgaon Own 100 USA,UK based Companies.

Sellcraft Softech Pvt. Ltd. eLearning, Document Management, Inbound / Outbound voice, Contact Center / Customer Care, Chat and Email Support Pune.

Service1ne Global Business Solutions Residential & Commercial Loan Servicing, Loan/Mortgage Processing Ahmedabad, Delhi, Schaumburg (US) 1000 Mortgage brokers, Mortgage bankers and Banks.

Seyyone Software Solutions Pvt. Ltd.. Medical Transcriptions Coimbatore.

SGRJ & Associates Accounting & Finance, Data Entry, Process Outsourcing, Legal Outsourcing, Legal Research Delhi.

Shubh Acccounting Solutions Pvt. Ltd. Claims Processing, Accounting & Finance, Payroll, Corporate Secretarial Services, Loan/Mortgage Processing, Legal Outsourcing Vadodara Internal Funding 25 Due to NDA entered with the clients, names are not disclosed.

Silgate Solutions Pvt. Ltd. Accounting, Tele-Marketing, Insurance Claims Processing, Collection Services, Upselling and Cross-selling, Medical Billing Mumbai.

Silicon Softech India Ltd. Tele-Marketing, Inbound / Outbound voice, Contact Center / Customer Care, Collection Services, Lead Generation Navi Mumbai.

Simple Soft Solutions Management Inc Data Processing / Management, Technical Support / IT Helpdesk, Accounting & Finance, Inbound / Outbound voice, Litigation, Healthcare Claims Processing Chennai, Hyderabad, USA.

Simtech Solutions Inc. Content Management, Email Support, Surveys, Data Entry, Lead Generation, Third Party QC/ QA Bangalore.

SixthElement Systems (P) Limited Market Research, Surveys, Lead Generation, Data Validation Services, Presales, Inside Sales Hyderabad.

SkillMeter People Solutions, HR, Employer Services Bangalore.

SMARTBridge Solutions Pvt. Ltd. CRM, Accounting & Finance, Knowledge Management, Transaction Processing Hyderabad – 25

SMR Solutions Pvt. Ltd. Technical Support / IT Helpdesk,

Network and Security Management, Remote Network Monitoring New Delhi Private 69 Small & Medium ISPs in US.

SMTELE Sys Ltd. Tele-Marketing, Data Entry, Chat and Email Support New Delhi, Noida - -

sNiche InfoTech Private Limited. Inbound / Outbound voice, B2B Lead Generation, Contact Center / Customer Care, Insurance, Finance, Presales Bangalore.

Softcall Services Inbound / Outbound voice Mumbai 250 US based clients.

Soft InfoTech, Inc. Accounting, Data Processing / Management, Data Conversion, Accounts Receivable / Payable / Reconciliation, Tax Forms, Tax Management Chennai - India.

Software Nucleus Inc Advanced Tech Support Hyderabad, Chicago.

SPARK Technologies CRM, Data Conversion, XML Tagging, e-publishing, Data Mining Chennai.

Spheres Medical Transcriptions Bangalore, Coimbatore 1200 Major hospitals and medical clinics in the USA.

Sri Chaitanya Infosystems Ltd. Technical Support / IT Helpdesk, Tele-Marketing, Collection Services, Data Services Hyderabad Private 250 Major Telecom Vendors, Major US Finance & Insurance Companies, Major UK Medicals.

Sri Sai InfoTech Data Processing / Management, Data Conversion, Data Entry, Third Party QC/QA Bangalore Indian Clients 25 south western railway, Datascript, iwebresources, udhagam infotech, Prognosys Technologies, Renuka associates, Bapuji Medical college, etc. overseas clients: Techcircle Inc.

SRV Technologies Content Management, Data Capture & Processing, Animation, Data Conversion, Data Entry, OCR Services Kolkata.

SSPS (P) Ltd. Medical Transcriptions, Inbound / Outbound voice, Inbound Call Center, Medical Coding Services, Medical Billing Hyderabad .

Staffing Solutions (India) Pvt. Ltd. Claims Processing, Retirement Planning New Delhi.

Stalwart Equity Research Services Offshore Advisory Services, Investment Advisory Pune.

Starry Datamation CRM, Data Conversion, Data Mining Kanpur.

Steering Jobs Pvt. Ltd. HR, Resume Management Services, Staffing Solutions and Training Bangalore.

STP Software India private Limited Insurance Claims Processing, Data Digitization, XML Tagging, e-publishing, Litigation, Healthcare Claims Processing Pune.

Strategic Solutions Pvt. Ltd. Data Processing / Management, Network and Security Management, Accounts Receivable / Payable / Reconciliation, Payroll, Tax Forms, Process Outsourcing Ahmedabad Can be submitted latter on. 150 Clients from USA,UK and Australia.

Strategix, Inc Inbound / Outbound voice, Credit Card Processing, Finance, Healthcare Claims Processing, Loan/ Mortgage Processing Dayton (US), Ahmedabad.

Stride Services Data Processing / Management, Insurance Claims Processing, Medical Transcriptions Delhi.

Subhashree Infotech Technical Support / IT Helpdesk, HR, Supply Chain Services, Data Mining, Chat and Email Support, Logistics Tracking Chennai.

Subset Services Data Capture & Processing, Accounting & Finance, Market Research, Data Entry New Delhi.

Summit HR Worldwide HR, Payroll, Recruitment Services Chennai Greenwich Venture Capital Partners 145 Sun

Microsystems, Cisco, SLAC, Lockheed

Sun InfoTech Data Capture & Processing, Data Conversion, Data Entry, e-publishing, Document Imaging, Data Services Chennai Privately owned 100 AEL Data.

Sunarc Technologies Pvt. Ltd. Data Processing / Management, Document Management, Data Entry, Data Mining, Resume Management Services Bikaner.

Sundaram Business Services Accounting & Finance, Accounts Receivable / Payable / Reconciliation, Insurance, Payroll, Benefit Administration Services (PF/ESIC), Check Processing Chennai(2).

Suntec Web Services Pvt. Ltd. Data Capture & Processing, Data Entry, XML Tagging, Business Services Provision, Content Outsourcing Services New Delhi.

SureVin Internet Services Ltd. Tele-Marketing, Collection Services Bhopal Promters funding 70 Reliance Infocomm Ltd,Tatateleservices Ltd,BTA Cellcom Ltd.

Sutherland Global Services Technical Support / IT Helpdesk, B2B Lead Generation, Contact Center / Customer Care Chennai, Mumbai, USA, Canada, Philippines.

SwarajComm Technologies Pvt. Ltd. eLearning, Technical Support / IT Helpdesk, Network and Security Management, Telecom Sector, Remote Network Monitoring, Data Services Hyderabad 22 Indosoft Enterprizes, Intergraph Shanta Software Pvt Ltd, JP Mobile, South Central Railways, Jet King

Swift Response Tele-Marketing, Email Support, Chat Support, Inbound Call Center Chennai, Hyderabad Angel Funding 350

Sykes Technical Support / IT Helpdesk, Marketing Program Support Amsterdam, Bangalore, Florida, Manila.

Symphony Data Corporation Healthcare Claims

Processing, Billing Collections Madison, Milwaukee, Phoenix, Hyderabad.

Tandem Consulting Pvt. Ltd. Accounting & Finance, Accounts Receivable / Payable / Reconciliation, Payroll, Tax Forms, Statutory Compliance, Benefit Administration Services (PF/ESIC) Chennai.

Teamworks Softech Pvt. Ltd. eLearning, Animation, e-publishing, Business Services Provision, Marketing Program Support Hyderabad.

Tech-Net Scientific e-Publishing Services e-Learning, e-publishing, Litigation Chennai, Hyderabad, USA.

Tech-Synergy Data Processing / Management, Business Transcription Services, Accounting & Finance, Contact Center / Customer Care, Transaction Processing, Financial Transcription Indore.

TechBooks Content Outsourcing Services New Delhi & US - 2100 US and European based publishers, information aggregators, professional societies, government agencies and major corporations.

TekLitigation Data Capture & Processing, Data Conversion, OCR Services, e-publishing, Litigation, Document Imaging Chennai.

Tele Sourcing Billing, Contact Center / Customer Care, Healthcare Payer, Claims Administration, Member Management Pune.

Telegenisys Pvt. Ltd. Accounting, Technical Support / IT Helpdesk, Billing Pune Private 200 US, UK & Canada based companies.

Telnet Telecom India Pvt. Ltd. Inbound / Outbound voice, Telecom Sector Bangalore 100 % Subsidary of International Telnet Inc., USA 55 US and UK corporations.

Tera Link Systems Pvt. Ltd. Accounting & Finance, Accounts Receivable / Payable / Reconciliation, Payroll, Tax Forms Chennai.

Thathwamasy Infotec Form Processing, Data Processing / Management, Data Capture & Processing, Data Conversion, Data Entry, e-publishing Kochi Private Limited Company 500 U.K/ Canada Based.

The Document Factory BPO Content Management, Document Management, Data Digitization, e-publishing, Data Mining, Content Outsourcing Services Mumbai 100% subsidary of Document Factory Group 250 Leading Publications, Universities, News Paper Groups across North America, UK.

Think Support Solutions Technical Support / IT Helpdesk, Contact Center / Customer Care, Helpdesk Services, Chat and Email Support, Advanced Tech Support Nasik NA 30 NA.

Thrinaina Informatics Ltd. Data Digitization, OCR Services Hyderabad.

TopSource InfoTech Solutions Data Capture & Processing, Business Transcription Services, Accounting & Finance, Contact Center / Customer Care, Accounts Receivable / Payable / Reconciliation, Payroll, Admin, Lead Generation, Recruitment Services, Process Outsourcing Pune, London.

Topsys Solutions Pvt Ltd Accounting, Accounting & Finance, Accounts Receivable / Payable / Reconciliation, Payroll, Tax Forms, Periodical returns (GST,VAT) reporting Bangalore.

Toyo Computers Pvt. Ltd. Data Capture & Processing, Data Conversion, Data Entry, Data Digitization, Engineering Services, CAD/GIS Mumbai.

Trac Mail Technical Support / IT Helpdesk, Tele-Marketing, Email Support, Chat Support, Order Processing, Collection Services, Helpdesk Services Mumbai View Group, eTec Ventures, Solectron-Stream 2000 Fortune 500 customers in

Retail, IT, Banking, Telecom.

Trac Technologies Data Processing / Management, Data Capture & Processing, Data Conversion, Data Entry, Data Digitization, e-publishing Pune.

Tracobi Healthcare Services Medical Transcriptions, Medical Coding Services, Medical Billing Hyderabad.

Tranquilmoney, Inc. Data Capture & Processing, Accounts Receivable / Payable / Reconciliation, Structured Finance (Mortgage/Asset backed Securities), Healthcare Claims Processing, Medical Billing New York, Chennai.

TransForm Solution Pvt. Ltd. Form Processing, Claims Processing, Data Capture & Processing, Credit Card Processing, Data Conversion, Document Imaging Surat.

Transfront Solutions Data Processing / Management, Claims Processing, Medical Transcriptions, Inbound / Outbound voice, Contact Center / Customer Care Calcutta.

Ultima Data Systems Data Processing / Management, Data Capture & Processing, Data Conversion, Data Entry, Data Digitization, Data Mining Mumbai Private 40 Witsun Technologies Pvt Ltd, Expand Technologies, Palak Communication Pvt LTd.

Uniiverse Soft Tech Form Processing, Data Capture & Processing, Data Entry, OCR Services, Data Mining Coimbatore.

UnilinkTeleservices (I) Pvt. Ltd. Inbound / Outbound voice, Contact Center / Customer Care, Lead Generation, Healthcare Payer Mumbai A call center with a capacity of 120 agents.

Unisoft Accounting Services Accounting, Data Processing / Management, Accounting & Finance, Data Entry Ahmedabad Accounting, Bookeeping Services.

Unisoft Datatech Data Processing / Management, Data Capture & Processing, Data Conversion, Data Entry, OCR

Services Ahmedabad 15 USA, UK and JAPAN

United Sourcing, INC. Tele-Marketing, Claims Processing, Contact Center / Customer Care, Collection Services, Residential & Commercial Loan Servicing, Medical Billing California, USA, New Delhi, Bangalore Private 200 Mortgage Banks, Healthcare: Medical Billing and Coding Companies, Collection Agencies

Universal Computer Consultancy (Galagali Group) eLearning, Data Capture & Processing, Animation, Data Entry, XML Tagging, e-publishing Mumbai.

Universal Cyber Info way Ltd - Kalldesk Tele-Marketing, Inbound / Outbound voice, Contact Center / Customer Care, Lead Generation Chandigarh.

Universal Technologies Medical Transcriptions Bhubaneswar.

Vaarisoft technologies Technical Support / IT Helpdesk, CRM, B2B Lead Generation, Banking, Business Services Provision, Advanced Tech Support Chennai.

Value Notes Database Pvt. Ltd. Content Management, Market Research, Competitor Analysis Pune.

VCSinfotech Data Processing / Management, Data Capture & Processing, Data Entry, Third Party QC/QA Chennai Funded by Directors 40 Global Enterpreneurs,Blue soft Technologies,Aubaa,Divinity Dice.

vCustomer Services India Pvt. Ltd. Technical Support / IT Helpdesk, Inbound Sales, Transaction Processing, Upselling and Cross-selling, Chat and Email Support New Delhi (3), Pune Oki Development, WestRiver Capital and Warburg Pincus 3500 Hi Tech, Retail, Hospitality Clients.

Vee Insure Form Processing, Check Processing, Adjudication of Claims, Claims Administration Bangalore Private

1000 Fortune 500 clients

Vee Technologies Claims Processing, Insurance Claims Processing, Billing, Medical Coding Services, Medical Billing Bangalore.

veHere Interactive Data Processing / Management, Inbound / Outbound voice, Accounts Receivable / Payable / Reconciliation, Market Analytics Kolkata $100,000 20 UK Consultancy MNC, A MNC Audit Firm.

Velan Info Services Technical Support / IT Helpdesk, Email Support, Helpdesk Services Coimbatore.

VServe Solution Form Processing, Data Processing / Management, Claims Processing, Market Research, Data Mining, Document Imaging Coimbatore.

Webahn Internet Solutions Pvt. Ltd. Medical Transcriptions Trivandrum subsidiary of Webahn Inc., USA 45 US Hospitals, Clinics.

Webden Form Processing, Data Conversion, e-publishing, Litigation Chennai.

Webshastra Pvt. Ltd. Email Support, Customer Acquisition, Lead Generation, Customer Interaction Management, Marketing Program Support Bangalore.

Westfort Tech Q Solutions Form Processing, Data Conversion, Data Digitization, OCR Services, e-publishing, CAD/ GIS Trivandrum.

Wizie.COM Data Processing / Management, Contact Center / Customer Care, Airfare Filling & Pricing Support Guragon, Michigan (USA).

WNS Global Services Pvt. Ltd. Accounting & Finance, Knowledge Management, Telecom Sector, Health Care, Travel, Transportation, Benefit Administration Services (PF/ESIC), Airline, Customer Interaction Management, Marketing Program

Support Mumbai, Pune, Gurgaon & Ipswich (UK) British Airways and Warburg Pincus (majority stake holder) 9000 British Airways, Royal&Sun Alliance, WPP Group, Federal Express, SITA

World Link (INDIA) Contact Center Pvt. Ltd. Inbound / Outbound voice, Data Conversion, Data Entry, Customer Management, Insurance Administration BANGALORE OWN FUNDS 150 3G UK, US mortgage, Southern Cross Telco Australia, advanced engine optimization Philipines.

WOX Systems Data Capture & Processing, Business Transcription Services, Data Conversion, OCR Services, XML Tagging, e-publishing Coimbatore Own Funding 50 Indian and International Clients

Xicom Technologies Pvt. Ltd. Data Capture & Processing, CRM, Accounting & Finance, Inbound / Outbound voice, Chat and Email Support, Advanced Tech Support New Delhi.

Xplore-Tech Services Private Limited. Tele-Marketing, Inbound / Outbound voice, Contact Center / Customer Care, Customer Acquisition, Telecom Sector, Billing Collections KOLKATA OWN 600 Telecom - Landline, Mobilephone & Broadband, Entertainment - Direct to home satellite, television service, Financial - Mortgage processing, Healthcare - Medical Insurance, Telecom Surveys Internet - websites, isp, broadband phones, business directories

Xtelesis Datacom Pvt. Ltd. Network and Security Management, Remote Network Monitoring Bangalore.

Zecon InfoTech Form Processing, eLearning, Claims Processing, Data Conversion, Data Entry, Data Mining Mumbai.

Zenta Group Inbound / Outbound voice, Transaction Processing Mumbai - 1700 Fortune 1000 companies in the US and UK.

Zeus Learning eLearning, XML Tagging, e-publishing, Content Outsourcing Services Mumbai, India Funded through internal accruals. 50 Publishing companies, eLearning companies and other education related companies in US and India including Fortune 100 companies.

Zodiant technologies Data Capture & Processing, Insurance Claims Processing, Data Conversion, Medical Coding Services, Healthcare Claims Processing, Medical Billing chennai.

Zuegma Software Tech Pvt Ltd Form Processing, Data Processing / Management, Data Conversion, Data Entry, e-publishing.

××××××

20. How to Write & Speak Better English 95/-

21. Quote Unquote (A Handbook of Quotations) 125/-

22. Improve Your Vocabulary 125/-

23. Common Errors in English 95/-

24. The Art of Effective Letter Writing 95/-

25. Synonyms & Antonyms 95/-

26. Idioms 95/-

27. Business Letters 90/-

Lotus PRESS

4263/3, Ansari Road,
Darya Ganj, New Delhi-110 002
Ph.: 32903912, 23280047,9811594448
E-mail: lotus_press@sify.com
www.lotuspress.co.in